Topics in
Current Physics

29

Topics in Current Physics

Founded by Helmut K. V. Lotsch

1 **Beam-Foil Spectroscopy**
Editor: S. Bashkin

2 **Modern Three-Hadron Physics**
Editor: A. W. Thomas

3 **Dynamics of Solids and Liquids by Neutron Scattering**
Editors: S. W. Lovesey and T. Springer

4 **Electron Spectroscopy for Surface Analysis**
Editor: H. Ibach

5 **Structure and Collisions of Ions and Atoms**
Editor: I. A. Sellin

6 **Neutron Diffraction**
Editor: H. Dachs

7 **Monte Carlo Methods** in Statistical Physics
Editor: K. Binder

8 **Ocean Acoustics**
Editor: J. A. DeSanto

9 **Inverse Source Problems** in Optics
Editor: H. P. Baltes

10 **Synchrotron Radiation**
Techniques and Applications
Editor: C. Kunz

11 **Raman Spectroscopy** of Gases and Liquids
Editor: A. Weber

12 **Positrons in Solids**
Editor: P. Hautojärvi

13 **Computer Processing of Electron Microscope Images**
Editor: P. W. Hawkes

14 **Excitons** Editor: K. Cho

15 Physics of **Superionic Conductors**
Editor: M. B. Salamon

16 **Aerosol Microphysics I**
Particle Interactions
Editor: W. H. Marlow

17 **Solitons**
Editors: R. Bullough, P. Caudrey

18 **Magnetic Electron Lenses**
Editor: P. W. Hawkes

19 Theory of **Chemisorption**
Editor: J. R. Smith

20 **Inverse Scattering Problems** in Optics
Editor: H. P. Baltes

21 **Coherent Nonlinear Optics**
Recent Advances
Editors: M. S. Feld and V. S. Letokhov

22 **Electromagnetic Theory of Gratings**
Editor: R. Petit

23 **Structural Phase Transitions I**
Editors: K. A. Müller and H. Thomas

24 **Amorphous Solids**
Low-Temperature Properties
Editor: W. A. Phillips

25 **Mössbauer Spectroscopy II**
The Exotic Side of the Method
Editor: U. Gonser

26 **Crystal Cohesion and Conformational Energies** Editor: R. M. Metzger

27 **Dissipative Systems in Quantum Optics**
Resonance Fluorescence, Optical Bistability, Superfluorescence
Editor: R. Bonifacio

28 **The Stratospheric Aerosol Layer**
Editor: R. C. Whitten

29 **Aerosol Microphysics II**
Chemical Physics of Microparticles
Editor: W. H. Marlow

30 **Real Space Renormalization**
Editors: J. M. van Leeuwen and T. W. Burkhardt

31 **Hyperfine Interactions of Radioactive Nuclei**
Editor: J. Christiansen

32 **Superconductivity in Ternary Compounds I**
Structural, Electronic, and Lattice Properties
Editors: O. Fischer and M. B. Maple

33 **Molecular Collision Dynamics**
Editor: J. M. Bowman

34 **Superconductivity in Ternary Compounds II**
Editors: M. B. Maple and O. Fischer

Aerosol Microphysics II

Chemical Physics of Microparticles

Edited by W. H. Marlow

With Contributions by
H. P. Baltes I. P. Batra B. J. Berne W. H. Marlow
R. V. Mikkilineni E. Šimànek P. E. Wagner

With 50 Figures

Springer-Verlag Berlin Heidelberg New York 1982

Dr. William H. Marlow

Department of Energy and Environment; Brookhaven National Laboratories Associated Universities, Inc., Upton, NY 11973, USA

ISBN-13:978-3-642-81807-3 e-ISBN-13:978-3-642-81805-9
DOI: 10.1007/978-3-642-81805-9

Library of Congress Cataloging in Publication Data. Main entry under title: Aerosol microphysics II. (Topics in current physics ; 29) Includes bibliographical references and index. 1. Aerosols. 2. Particles. I. Marlow, W. H. (William H.) II. Baltes, Heinrich P. III. Series. QC882.A352 541.3'4515 82-830 AACR2

2153/3130-543210

Preface

Aerosols, which are gas-phase dispersions of particulate matter, draw upon and contribute to multidisciplinary work in technology and the natural sciences. As has been true throughout the history of science with other fields of interest whose underlying disciplinary structure was either unclear or insufficiently well developed to contribute effectively to those fields, "aerosol science" has developed its own methods and lore somewhat sequestered from the main lines of contemporary physical thought. Indeed, this independent development is the essential step in which systematic or phenomenological descriptions are evolved with validity of sufficient generality to suggest the potential for development of a physically rigorous and generalizable body of knowledge. At the same time, the field has stimulated many questions which, limited to its own resources, are hopelessly beyond explanation. As Kuhn pointed out in *The Structure of Scientific Revolution* [2nd enlarged edition (University of Chicago Press, Chicago 1970) Chapter II and Postscript-1969] this is a very common juncture in the development of a science. In brief, the transition from this earlier stage to the mature stage of the science involves a general recognition and agreement of what the foundations of the field consist of. By this critical step, a field settles upon a common language which is well defined rather than the ambiguous, and often undefined descriptors prevalent at the earlier stage. These volumes on aerosol microphysics propose part of a candidate approach to the physical foundations of aerosol science which is biased toward maximizing its connections with basic fields of physical research.

To a large extent, the transition from phenomenological to physically rigorous work in aerosols has been in progress for some time (a confusing state of affairs Kuhn also recognized as commonplace in the transition). Unavoidably, misunderstanding on the parts of both the new contributors from the basic scientific disciplines and the practitioners of the traditional phenomenological approaches arise. For the new contributor, a tendency not to appreciate the importance of the influences of other areas than his own upon the aerosol behavior he seeks to explain is a danger. For the traditionally oriented aerosol investigator, the complementary problem arises. He may see the new classification into reasonably well defined and distinct categories from physics, chemistry, and basic engineering science as simple reductionism that ignores the complexity of the field. Indeed, this is reductionism, but it is one which illuminates the complexities and recognizes that the tasks of their integration in order to describe real aerosols constitute the important categories of aerosol macrophysics which are not addressed in these volumes.

As conceived in these volumes, aerosol microphysics is the study of one or a small number of particles in a gas of one or a small number of molecules. It consists of a number of fields among which no natural division exists, except for editorial convenience. This editor proposes those fields to include the kinetic theory of particles in a gas, optical interactions, particle interaction forces, heterogeneous interactions of molecules with particles, homogeneous nucleation, microparticle physics, and aerosol thermodynmamics. Due to the fact that the subdivisions of aerosol microphysics center largely on a single particle, they are usually incapable of providing the explanation for *aerosol* phenomenon. However, since this is the level at which aerosols become specific physical and chemical entities, it is also the level at which aerosol science must cease to be phenomenological or qualitative and at which it must incorporate the physical and chemical properties of its constituents. Since these subdivisions are important fields on their own, their independent advancement contributes directly to the rigorous development of aerosol microphysics.

Aerosol macrophysics are the studies of aerosols as systems. Distributions of particles occur in numerous questions such as optical turbidity, heterogeneous gas-phase chemistry, electrostatic charging and migration, filtration, and clouds. Consequently, aerosol macrophysics must integrate large-scale systems' descriptions with the characteristics of the microphysical behavior of the particles that comprise the aerosol.

*Aerosol Microphysics I** and *II* present a series of essays on various aspects of the subject. Many of the contributors know little if anything about aerosols but are distinguished in their own fields. The perspectives they give will hopefully stimulate aerosol investigators to broaden their own approaches and will interest contributors from their several fields in the challenges and opportunities offered by the aerosol condition of matter.

Upton, April 1982 *William H. Marlow*

---

*Published as Volume 16 of Topics in Current Physics

Contents

List of Contributors

Baltes, Heinrich P.

LGZ Landis and Gyr Zug AG, Zentrale Forschung und Entwicklung,
CH-6301 Zug, Switzerland

Batra, Inder P.

IBM Research Laboratory K33-281, 5600 Cottle Road, San Jose, CA 95193, USA

Berne, Bruce J.

Havemeyer Hall, Department of Chemistry, Columbia University,
New York, NY 10027, USA

Marlow, William H.

Department of Energy and Environment, Brookhaven National Laboratory,
Associated Universities, Inc., Upton, NY 11973, USA

Mikkilineni, Rao V.

HO2F 412, Crawford's Corner Road, Bell Laboratories,
Holmdel, NJ 07733, USA

Šimànek, E.

Department of Physics, University of California, Riverside, CA 92521, USA

Wagner, Paul E.

Institut für Experimentalphysik, Universität Wien, Strudlhofgasse 4,
A-1090 Wien, Austria

1. Aerosol Chemical Physics

W. H. Marlow

1.1 Aerosol Microphysics

The presence of gas-phase dispersions of particulate matter in diverse natural and man-made systems as well as the existence of common body of experimental and computational methods for dealing with them justify the denomination *aerosol* for all such dispersions [Ref.1.1, Chap.1]. The computational aerosol methods that are largely qualitative today cannot become quantitative and specific until they are capable of incorporating the chemical and physical characteristics of each system or until the conditions for the appropriate omission of such details can be rigorously established. Specific chemistry and physics usually involves the formation or interaction processes of one or a small number of particles. Thus, *aerosol microphysics* is the name given to this subdivision of aerosol physics [Ref.1.1, Chap.1] that treats one or a small number of particles in a gas. In this scheme an important part of *aerosol macrophysics*, the complementary subdivision of aerosol physics, is the integration of microphysical results to give a description of the aerosol as a whole.

If a scheme for the classification of research efforts is to be of use as a guide to the description of aerosols, it must identify the links between questions specific to aerosols and general areas of basic physical science. The divisions chosen in these books attempts to do this. Generally speaking, one can consider a range of questions involving the aerosol particles wherein their composition is unchanged or they undergo interactions which permit parameterizations that are only implicitly dependent upon composition. This part of aerosol microphysics constitutes Volume I [1.1] and may be taken to include the kinetic theory of aerosols (including transport and molecular accomodation), optical interactions, and particle interaction forces. The present volume addresses those other questions in which specific chemical and physical characteristics are incorporated. In general terms, these areas are the physics of clusters and very small particles, heterogeneous (or surface) processes, homogeneous nucleation, and aerosol thermodynamics. Of course, such a breakdown is, to a large extent, arbitrary both due to the size and compositional dependences of the parameterizations and due to the significant interdependences among the divisions of aerosol microphysics. Some of these links will be mentioned in the following discussion.

1.2 Chemical Physics of Microparticles

1.2.1 Isolated Particles and Clusters

The component of aerosols which distinguishes them from the purely gas phase is the particle or cluster. If these nonmolecular entities all were spheres that differed only in radius and otherwise had no chemical or physical interactions, then aerosol science in such a fictitious world would be solely a complicated branch of fluid mechanics. The fact that aerosols interact and react in specific, compositionally dependent ways dictates that these fundamental components of the aerosol be thoroughly characterized. The physics of microparticles therefore is an essential part of aerosol microphysics.

Microparticles and clusters constitute an emerging field of research of their own [1.2,3]. Because of their size, such particles have properties that differ from those of the same material in bulk. However, no distinct size can be identified below which all properties cease to be "bulklike". Rather, the size depends upon the property under investigation (Sect.2.1.1) much as transport properties do (Sect.5.3.1). The questions therefore arise as to what microparticle properties are of interest in aerosol work and how they differ from what would be expected from purely bulk-material properties. The answer is that most questions that can be asked about microparticle physics will likely be shown to be relevant to aerosols in one way or another. For example, one area of great importance is aerosol formation via nucleation processes. An essential part of current theory is the change in free energy of the nucleating species as it forms the cluster or particle from the gas phase [1.4]. Since the free energy of the gas phase is adequately described by available methods, the energetics of the microparticle is where the greatest uncertainty lies. This problem has several aspects because both the energy involved in forming the microparticle structure (commonly parameterized by particle surface tension) and the partitioning of excess energy among phonon and electronic modes as well as the small-particle radiation content may be involved. What marks the phonon and radiation mode questions as being of particular fundamental physical interest is that they are both associated with wavelengths. In microparticles, the wavelengths characteristic of the bulk materials are generally greater than the particle dimensions, which leads to the problem of the eigenmodes of a finite body (Sect.2.2). In addition to excess-energy partitioning questions, small-cluster structure is unlikely to reflect bulk structure in such a manner that the bulk surface tension will meaningfully represent the cluster's "surface tension". Thus, recourse to formal definitions of surface tension (Sect.4.3.1) is required if the highly successful, though somewhat phenomenological, correlations of classical nucleation theory [1.5] are to be better understood and used as guides to a more fundamental theory. The question of the thermodynamics of small systems (e.g., a small cluster) is also pertinent here.

In the classical theory of nucleation and elsewhere in aerosol studies, little attention has been given to the necessity of differentiating the thermodynamic properties of the gas, which are bulk quantities, from those of the particle. This differentiation is sometimes acknowledged in the recognition of a probable breakdown, with diminishing particle size, in the Kelvin relation. This relation gives the elevation of the vapor pressure over a small spherical particle relative to that over a plane surface based upon macroscopic surface tension. Since it depends upon a macroscopic thermodynamical model for the particle, it must break down for sufficiently small particles. An analogous breakdown occurs in the heat capacity of a particle in the nanometer size range. For a particle of that size, the heat capacity increases more slowly with temperature at low temperatures than the bulk Debye value due to the substantial fraction of atoms on the particle surfaces (Sect.2.2.2).

Another area where microparticle physics is of great importance is in the thermal radiative emissions of small particles. In physical circumstances as diverse as engineering systems (e.g., combustors) and domains of astrophysical interest, questions of radiative transfer in particle-laden regions are important. This question involves finite-body radiation modes, heat content of small bodies, and distortion of emission characteristics due to finite particle size (Sect. 2.2.3). Infrared emission by microparticles may also be important for chemical analytical purposes. However for microcrystals a substantial complication can arise due to the scrambling of optical and acoustic modes leading to a spectrum different from the bulk material (Sect.2.3.1).

In [Ref.1.1,Chap.4], a classical model for inelastic light scattering by a dielectric sphere was presented. This model is an extension of the Mie theory which is parameterized in terms of a frequency-dependent complex dielectric constant. However, this dielectric constant is a function of the microparticle surface modes and these can be different from those predicted for bulk material either due to mode scrambling or other causes (Sect.2.3.2). Similarly, the Lifshitz-van der Waals interactions involving small particles [Ref.1.1, Chap.5] may also be affected.

Physical adsorption upon microparticles is an important part of the heterogeneous processes in which aerosol particles are involved. Electric and magnetic properties of the particles as a whole are important here and require much further exploration. As has been pointed out elsewhere [1.6], the deviation of metallic-particle electrical susceptibility from the bulk value for the metal has an effect on the particle Lifshitz-van der Waals interaction energy and presumably also upon its adhesion energy.

Aerosol methods, in principle, should also be of value in microparticle studies. One significant source of data is matrix-isolated samples. The question always arises as to what are the contributions of the matrix to the small particle data (Sect.2.4.6). Since aerosol particles are matrix-free, they may provide a less ambiguous experimental source of information via currently evolving measurement techniques (e.g. [1.7,8]) than would otherwise be available.

Contrary to physisorption, chemical adsorption likely involves only electronic properties of the particle surface. However, aerosol particles and clusters are often of the dimension of the surface thickness layer considered in surface studies. This leads to the picture of such particles as "three-dimensional" surfaces. Perhaps in no other property does the view of the aerosol as the transition condition of matter between the molecular and condensed phases show itself more clearly than in chemical bonding. The particle size dependence of this bond can be traced from the individual bonding of an atom to a monomer (ordinary chemical bond) through chemisorption (or surface chemical bonding) of the atom to a cluster and then a surface. This picture is made somewhat explicit in the calculational results given in Chap.3. There, the electronic density of states of a chemisorbed oxygen atom on aluminum clusters of five, nine, and twenty-five atoms are projected (Figs.3.2,3). Taking the density of states of the Fermi level as a crude indicator, these computations show that only for the largest cluster are there conduction electrons present as in bulk. This can be understood by referring to Fig.3.8 where calculational results for the electronic structure of oxygen adsorbed upon a plane aluminum substrate are given. There, an effectively occupied Fermi level does not come about until the second or third layer below the surface, which corresponds to the level of interior atoms present only in the largest cluster mentioned above.

1.2.2 Physical Transformations and Thermodynamics

Formation of thermodynamically stable clusters (or microparticles) in a gas free of foreign bodies (i.e., ions, walls) is a process that can occur by only one of two possible pathways for each type of molecule. Either cluster formation proceeds as an activationless process, in which case no evaporation from any cluster occurs, or a "barrier to nucleation" exists. This barrier is a manifestation of the unstable balance between vapor- and condensed-phase (i.e., cluster) free energies which results in a high probability of evaporation for clusters smaller than a thermodynamically determined "critical cluster" size and a high probability of growth for clusters larger than this size. This latter process, involving a barrier to nucleation is what is commonly referred to in the aerosol context as homogeneous nucleation as opposed to the energetically downhill (activationless) process of stable cluster formation by polymerization. The surprisingly successful and still state-of-the-art capillarity approximation in homogeneous nucleation theory is that critical cluster free energies (see [1.4] for a review) are based upon the assumption of macroscopic surface tension and bulk-phase chemical potential that cannot be correct for small enough clusters. The generality of the form of the classical theory's results for a very wide variety of substances has recently been shown by McGRAW [1.5], who presented extensive correlations of existing nucleation data by use of a corresponding states approach. In a sense, this similarity of behavior under these correlations for broad ranges of molecules suggests some common base for these trends in the currently unobservable subcritical cluster regime. Of course, for those the capillarity approximation will be less likely to hold. If this is true, then one would expect numerical simulations of simple systems to give some insight into these trends. Unfortunately, not all processes can be realistically modelled in this fashion. For example, a molecular dynamics simulation of the real time nucleation process under naturally occuring supersaturations would not appear feasible. However, important contributions to such questions as cluster structure and surface tension and the cluster-vapor interface do hold potential for contributing to the descriptions of the subcritical clusters. These methods have already shown themselves to be useful in the study of surfaces and are discussed in Chap.4.

In any system where a component of the gas phase is supersaturated, the potential for homogeneous nucleation exists if the supersaturation is sufficiently high. If heterogenieties such as aerosol particles are present, they deplete the gas phase of condensible vapor molecules and clusters thereby modifying the nucleation that can occur in one of two ways. In the first way, the nucleation rate is reduced as the steady state supersaturation is reduced through the removal of condensable vapor by the aerosol. If the time scale for this removal exceeds the time scale for nucleation, the coupling is that expected as a result of the depression of the vapor supersaturation ratio due to the presence of the aerosol. In this case, the classical nucleation rate expression with this corrected supersaturation ratio is applicable. Alternatively, when the time scales for cluster scavenging by the aersol and nucleation are comparable, a significantly modified version of the theory must be used [1.9]. This is true, for example, when the partial pressures of the condensing species are low and the lag time for nucleation long.

Condensational aerosol growth depends upon several areas of aerosol microphysics including both transport kinetics and aerosol thermodynamics. To fit experimental growth data for particles smaller than those of the diffusion-controlled regime, molecular accomodation coefficients have been invoked, while for larger particles such a parameter does not enter (Sect.4.3.2). Thus, for years there has been an apparent inconsistency that could only be traced to particle size relative to gas pressure. Since heat is evolved as molecules condense on the particle, a thermal gradient form the particle to the background gas temperature develops. The removal of this heat is materially different in the nondiffusion controlled regime from what it is for larger particles due to the discontinuity of the gas near the particle surface that permits a discontinuity in the temperature profile (Sect.5.3.3). When these discontinuities due to particle condensation thermodynamics are incorporated into the gas kinetic equations, the accomodation coefficients normally invoked may all be set to unity resulting in fits to the data that are superior to those derived by older methods.

References

1.1 W.H. Marlow (ed.): *Aerosol Microphysics, I: Particle Interaction*, Topics in Current Physics, Vol.16 (Springer, Berlin, Heidelberg, New York 1980)
1.2 Proceedings International Meeting on the Small Particles and Inorganic Clusters: J. Phys. Paris *38* Colloq C-2, Suppl. No.7 (1977)
1.3 Second International Meeting on the Small Particles and Inorganic Clusters: Surf. Sci. *106* (1981)
1.4 F.F. Abraham: *Homogeneous Nucleation Theory* (Academic, New York 1974)
1.5 McGraw: J. Chem. Phys. *75*, 5514 (1981)
1.6 W.H. Marlow: *Surface Sci. 106*, 529 (1981)
1.7 A.J. Campillo, H.-B. Lin: *Photo-opt. Instrum. Eng. 286*, 24 (1981)
1.8 A.J. Campillo, H.-B. Lin: Appl. Opt. *20*, 3100 (1981)
1.9 R. McGraw, W.H. Marlow: *Multi-State Kinetics of Nucleation in the Presence of an Aerosol*, J. Chem. Phys. (submitted)

2. Physics of Microparticles

H. P. Baltes and E. Šimànek

With 11 Figures

Microparticle physics is different from the physics of bulk matter. An impressive manifestation of this difference is the explosion of dust from materials that are quite harmless in the bulk state. Another conspicuous example is the blackness of colloidal gold. Microparticle physics is also distinguished from the physics of the surface of bulk media and the physics of thin films or whiskers. Far-infrared spectroscopy provides a striking example: bulk surfaces, thin films, and microparticles of alkali halides show quite different types of reflection or emission spectra. In bulk material, characteristic lengths such as the pertinent wavelengths, correlation lengths, and mean free paths are much smaller than the macroscopic geometrical dimensions of the sample; i.e., the sample boundary is not "seen" by the phenomenon in question. On the contrary, microparticles are solid samples whose geometrical dimensions are microscopic, *i.e.*, of the order of, or even smaller than, the microscopic characteristic length characterizing the physical effect under consideration, and thus the boundary plays an important, if not dominant, role.

This chapter presents exemplary aspects of microparticle physics as opposed to the physics of bulk media or macroscopic samples. It is not intended here to provide a full-scope review article covering all the relevant literature. Thus we shall refer only to a limited number out of the many exciting papers on this young, but rapidly growing field. Moreover, we shall not report any work on thin films, infinite slab geometry, or surfaces of large systems. In Sect.2.1 a definition of microparticles and their various size regimes is attempted. Moreover, a number of typical size effects are presented together with a condensed overview of the field and a guide to the review-type literature. In Sect.2.2 the emphasis is on pure boundary effects, i.e., on perfect gases or elementary excitations such as acoustical phonons in small enclosures with perfect boundaries and well-defined boundary conditions. Thermodynamic relations and the Bose-Einstein condensation for small systems are discussed. Photons and the pertinent radiation laws are considered as well. The related mathematical problem of finding the distribution of the eigenvalues of the wave equation for a finite domain (Weyl's problem) is summarized. Section 2.3 is devoted to optical phonons and the related far-infrared properties of dielectric microparticles. The electronic heat capacity and the electron spin susceptibility of metallic microparticles are discussed in Sect.2.4, along with the nuclear- and electron-spin resonance as well as the blocking of the spin relaxation processes and of the Kondo effect by the energy-level quantization. Section 2.5 is devoted to the electromagnetic properties such as electric polarizability, plasma resonance, and far-infrared absorption in metallic microparticles and their aggregates. Effects of superconductivity are considered in Sect.2.6 with emphasis on the thermodynamic fluctuations of the superconducting order parameter, which are responsible for the "rounding" of the phase transition in metallic microparticles. We discuss not only various ways of detecting these rounding effects, such as diamagnetic susceptibility, specific heat, ultrasonic attenuation and a nuclear spin-lattice relaxation, but also the modifications of the transition temperature caused by electronic and phonon size effects.

Before going into detailed discussion, it seems appropriate to point out that the comparison between experiment and theory in the young field of microparticles is usually not straightforward. Experiments often deal with an ensemble of microparticles of different shape and size and in contact with each other or some matrix or substrate material. Therefore existing resonance effects predicted by the theory for a perfect microparticle may be obscured in the experiment by the averaging over a broad size and shape distribution and the interaction with the matrix material. This situation is improved in experiments with beams of free particles or with single particles fixed in space by an electric field.

2.1 Introductory Remarks

Thermodynamics is usually understood as the study of the macroscopic properties of bulk systems [2.1], i.e., systems which have vanishing surface-to-volume ratios and include a very large number of atoms or molecules or elementary excitations. By definition the "specific" or "intensive" properties of bulk matter, such as specific heat and other thermodynamic derivatives or the electric or magnetic susceptibility, do not depend on the size and the shape of the sample. Microparticles, however, are characterized by a large surface-to-volume ratio and consist of a much smaller number of atoms. Microparticle physics can be understood as the study of size effects, i.e., the dependence of "specific" or "material" properties or other intensive parameters on the particle size.

2.1.1 How Small is Small?

From the point of view of the aerosol concept [2.2], microparticles can be defined as particles whose average diameter d is much smaller than 50 μm. This coarse upper limit is in agreement with current studies of microparticle solid-state physics, where the size of particles under consideration ranges between a few Å (corresponding to microclusters of a few atoms) and a few microns (corresponding to particles comparising as many as 10^{12} atoms or thereabout). Simple geometrical considerations show that the ratio of the number of surface atoms N_S to the total number of atoms N in a microparticle is of the order of $N_S/N \approx 4N^{-1/3}$, i.e., as large as 0.4 for $N \sim 10^3$, 0.2 for $N \sim 10^4$, and 0.1 for $N \sim 10^5$.

More specific definitions of the microparticle size depend on the particular physical property that is studied. For example, KCl particles of 10 μm average diameter are microparticles showing a strong size effect with respect to the far-infrared (FIR) spectroscopy, their size being much smaller than the wavelength $\lambda_{TO} \approx 70$ μm of the transverse optical phonon mode of the corresponding bulk material. On the other hand, 10-μm diameter samples may behave like bulk material with respect to other physical phenomena.

Quantum-size effects may be defined as the dependence of the properties of samples on their geometrical dimensions when the latter become comparable with the effective wavelength of the pertinent elementary excitation (e.g., phonons or electronic levels). In the quantum-size regime the particle size is so small that the distance Δ between adjacent eigenvalues of the elementary excitation energy is comparable with characteristic energies involved in the experiment under consideration, such as the thermal energy k_bT, the magnetic energy μ_BH, or some spectral resolution.

For example, a quantum-size effect in the vibrational specific heat is expected when the particle diameter d is so small and the temperature T so low that the difference between adjacent acoustical phonon levels, $\Delta \approx \hbar v/d$, is of the order of k_BT. Here v denotes the effective sound velocity and $\hbar$ Planck's constant divided by 2π. With respect to this quantum-size effect, microparticles may be defined as particles whose average diameter is $d \approx \hbar v/k_BT$ or smaller. With $v \approx 1$ km/s this leads to $d \approx 100$ Å/T[K]. The effect in question has indeed been observed in the low-temperature specific heat of sub-100-Å particles (Sects.2.1.2 and 2.2.2). Similarly, quantum-size modifications of the thermal radiation laws are expected in cavities of diameter $d \approx \hbar c/k_BT$, where c denotes the speed of light. This leads to $d \approx 3$ cm/T[K], i.e., the cavity diameter is of the order of the dominating wavelengths in Planck's distribution.

In general, microparticles can be defined only with respect to a given physical property as particles whose average diameter is of the order of (or at least not large compared with) a length characterizing the property under consideration. Beside the average thermal phonon and photon wavelengths considered above, we mention the de Broglie wavelength of electrons at the Fermi surface, the mean free path characterizing a transport property, and the correlation or coherence length describing a cooperative phenomenon such as superconductivity. It should be pointed out that the application of the above definitions to aerosol physics is not necessarily straightforward, since an aerosol is generally in a nonequilibrium state [5.2].

2.1.2 Exemplary Size Effects

We now present a brief cross section of the various phenomena of microparticle physics. We begin with the *far infrared (FIR) spectroscopy of optical phonons* in alkali halides. Figure 2.1 shows the thermal emission spectra of a bulk sample, a homogeneous thin slab, and a layer of microparticles on a (poorly emitting) metallic substrate. The microparticle spectrum differs strongly from not only the bulk, but also the thin-slab spectrum. While the bulk surface exhibits small emissivity

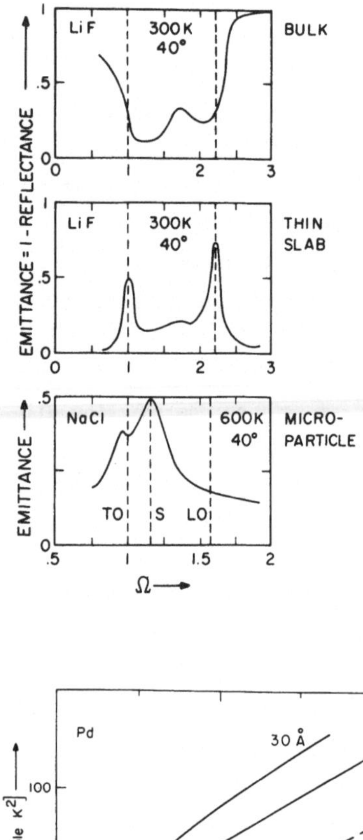

Fig. 2.1. Far-infrared emittance of alkali halide crystals as a function of the reduced frequency $\Omega = \omega/\omega_{TO}$. A layer of cube-shaped particles of average edge length of 2 μm on a metallic substrate (below) is compared with a thin slab of about 1 μm thickness (middle) and a bulk crystal (above). Materials, temperatures, average emission angles, as well as $\Omega_{LO} = \omega_{LO}/\omega_{TO}$ are indicated (adapted from [2.3])

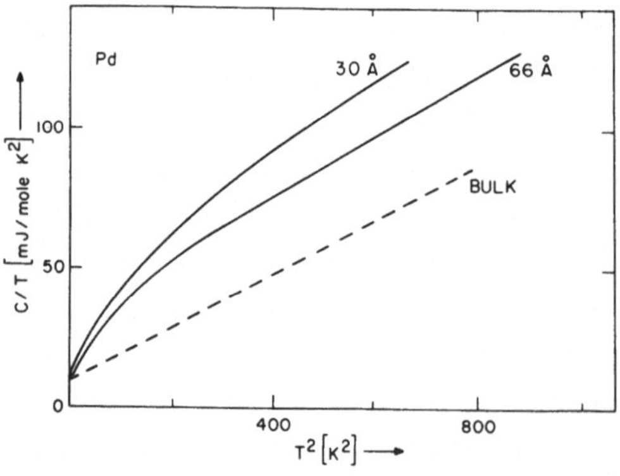

Fig. 2.2. Temperature dependence of the specific heat C of Pd microparticles in comparison with the bulk. Average diameters are indicated (adapted from [2.4])

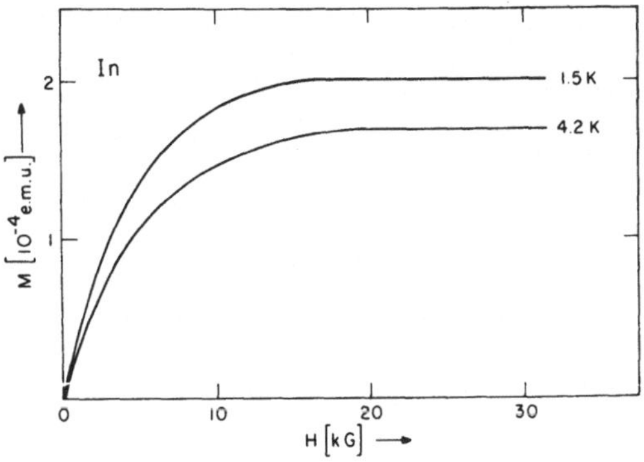

Fig. 2.3. Low-temperature magnetic moments of indium microparticles of 50 Å average diameter (adapted from [2.6])

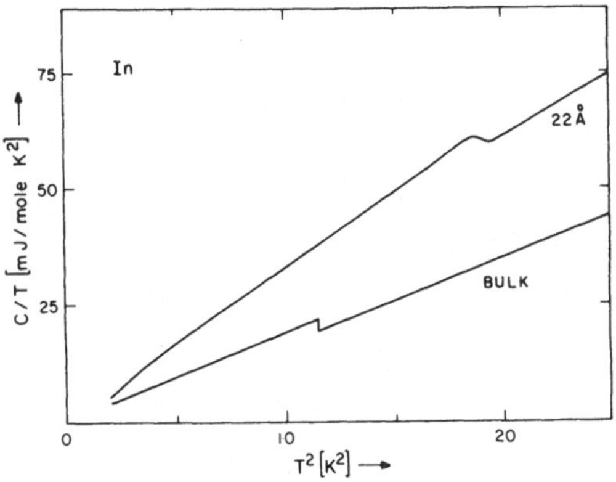

Fig. 2.4. Specific heat C of indium microparticles in comparison with the bulk in the neighborhood of the transition temperature of superconductivity (adapted from [2.7])

in the reststrahlen band between the bulk transverse and longitudinal optical mode frequencies ω_{TO} and ω_{LO}, the microparticles produce a strong emission band in this very spectral region. This band is peaked at some "surface mode" frequency ω_S which is missing in the slab spectrum. For further discussion we refer to Sect.2.3.

Next we consider the *low-temperature specific heat in the quantum-size regime*. Data of Pd microparticles of 30 and 60 Å average diameter (approximately 10^3 and 10^4 atoms) in comparison with the bulk specific heat are shown in Fig.2.2. The specific heat of the microparticles is strongly enhanced over the bulk, the enhancement being larger for smaller particles. Extrapolation to T = 0 in Fig.2.2 indicates that for palladium the electronic contribution to the specific heat is independent of the size in the investigated temperature range. The recorded size dependence is therefore ascribed entirely to the *vibrational* contribution. An enhancement of the vibrational specific heat is expected for free particles, since an appreciable part of the atoms are surface atoms (with weaker bonds) which lead to lower vibrational eigenfrequencies and hence to a larger low-frequency mode density. The results presented in Fig.2.2 can be described quantitatively in terms of a theoretical model (vibrational levels of homogeneous elastic spheres with free surface) proposed by BALTES and HILF [2.5]. For further discussion see Sect.2.2.

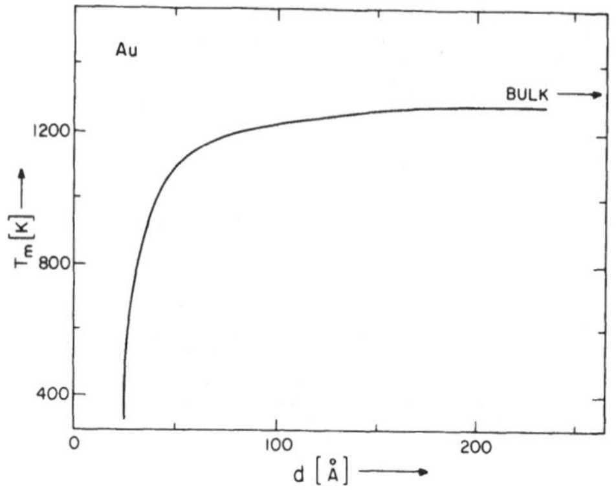

Fig. 2.5. Size dependence of the melting-point temperature T_m of gold particles (average diameter d)

MEIER and WYDER [2.6] have shown that an analogous model for the *electronic* levels of small metallic spheres can account for the *low-temperature magnetic moments* of indium microparticles of a diameter varying between 20 and 100 Å. These diameters correspond to average level distances between 100 and 2 K, i.e., in the quantum-size regime. Typical experimental results are shown in Fig.2.3.

Quantum-size phenomena involving both phonons and electrons are presumably responsible for the size dependence of the transition temperature T_c of *superconductivity*. This effect is demonstrated in Fig.2.4. The figure shows the specific heat of indium microparticles (22 Å average diameter) in comparison with bulk indium measured in the neighborhood of the transition temperature. Both a "smoothing" of the transition (Sect.2.6.3) and a shift of the transition temperature (Sect.2.6.6) are observed. Analogous effects may be expected for other second-order phase transitions.

Also the first-order phase transition of *melting* is subject to size effects. As an example we mention the size dependence of the melting temperature T_m of gold particles as measured by BUFFAT and BOREL [2.8]. Figure 2.5 shows the T_m for particles of an average diameter between 20 and 250 Å. The melting temperature is drastically lowered for diameters below about 50 Å. This behavior can be understood in terms of thermodynamic models [2.8].

The *crystallographic structure* of microparticles is often that of the corresponding bulk material with only modest modification of the lattice constant. For example, SOLLIARD and BUFFAT [2.9] report up to 1% contraction for gold particles of diameters between 140 and 25 Å. On the other hand, an interesting *crystallographic anomaly*, namely the formation of clusters with *pentagonal symmetry* (which cannot exist in any bulk material), is also observed (e.g., [2.10] and references therein). The particular structure realized, as well as the size distribution, seems to depend on the preparation method.

In view of the relevance of the *thermal radiation laws* to aerosol physics [2.2], we finally mention the case of the photon gas in thermal equilibrium in a small cavity. Figure 2.6 shows the total thermal energy E calculated for a lossless

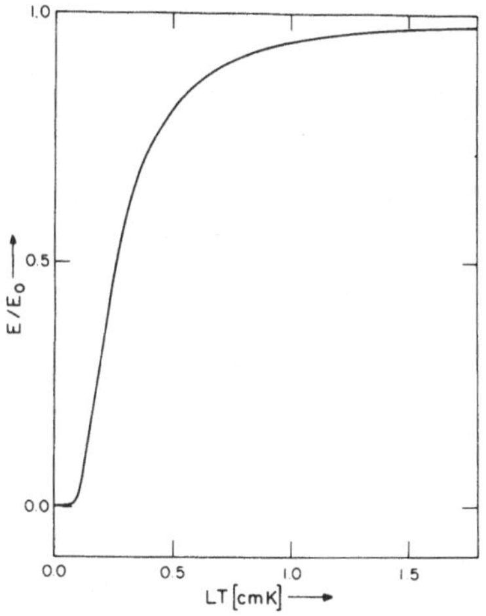

Fig. 2.6. Total thermal radiation energy E in a lossless cube-shaped cavity of edge length L at temperature T, normalized with respect to the Stefan-Boltzmann (bulk) limit E_0 (adapted from [2.11])

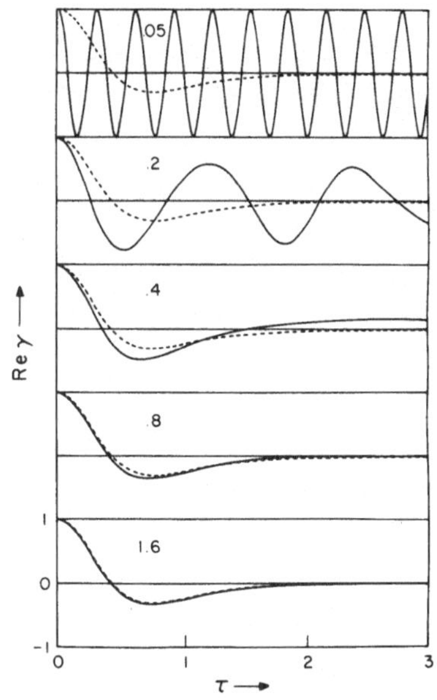

Fig. 2.7. Temporal electric field autocorrelation Re γ for thermal radiation in a lossless cube-shaped cavity of edge length L at temperature T for various values of LT between 0.05 and 1.6 cm K (full curves). The bulk limit (LT → ∞) is shown for comparison (dashed curves)

cube-shaped cavity of edge length L as a function of LT, where T denotes the temperature. The energy E is normalized with respect to the Stefan-Boltzmann value $E_0 \propto L^3T^4$ valid in the "bulk" limit of very large LT. In Fig.2.7 we show the corresponding temporal *autocorrelation* of the electric field, Re $\gamma(\tau)$, $\tau = tk_BT/\hbar$, with t denoting time. The complex function $\gamma(\tau)$ is essentially the Fourier transform of the spectral energy distribution [2.12]. We notice that the correlation is periodic (Poincaré-cycle), but the period remembering the existence of the cavity walls increases with increasing LT, i.e. increasing size. Such a size effect may occur for all kinds of correlations. Another example is the velocity autocorrelation of finite systems ([2.13] and references therein).

2.1.3 Overview and Guide to the Literature

The more general field of the physics of finite systems is not limited to the solid-state properties of microparticles but includes aspects of thermodynamics and statistical physics, quantum electrodynamics, and even nuclear physics, the acoustics of rooms, and astrophysics (interstellar grains). A tentative list of related problems of current interest may read as follows (see [2.3] for a more extensive list with references).

a) *Non-relativistic perfect quantum gases*. Thermodynamic equation of state. Surface and shape tension. Quasi-continuum versus quantum-size density of states. Bose-Einstein condensation. Poincaré cycles. Time and space correlations.

b) *Quantum electrodynamics*. Photon gas. Thermal radiation laws in small cavity. Fluorescence lifetime and Lambshift. Radiative heat transfer between closely spaced bodies. Casimir effect.

c) *Heavy nuclei in terms of independent particle model*. Nuclear masses and fission barrier. Energy-level statistics. Surface and shape energy. Shell correction.

d) *Solid-state thermodynamics*. Vapor pressure and melting temperature of microparticles. Phase transitions in the presence of boundaries. Zero-dimensional superconductor.

e) *Phonons in microparticles*. Acoustical phonons: lattice dynamics of surface modes, non-Debye specific heat, Mössbauer studies, neutron scattering, relevance for superconductivity. Gap modes. Optical phonons: FIR and Raman spectroscopy, optical surface modes.

f) *Electrons in microparticles*. Specific heat and paramagnetism in metallic microparticles. Electric polarization. Optical, UV, and FIR absorption, spin resonance. Tunneling. Statistics of electronic energy levels.

We conclude the introductory section by providing a *guide to the review-type literature* that may serve as a starting point for further reading. To the best of our knowledge, there is no book, neither monograph nor multiauthor volume, that fully covers the field of microparticle physics. The book by BALTES and HILF [2.14] on *Spectra of Finite Systems* is a rather mathematical state-of-the-art review focussed on Weyl's problem of determining the eigenvalue distribution of the wave equation for finite domains and provides only brief discussions of the various physical applications. Perhaps the best source of information still is the proceedings of the first international conferences on microparticles held in Lyon in 1976 [2.15]. Besides many original papers, it contains reviews on the geometries of soft-sphere packing by BARKER [2.16], on the discreteness of energy levels in small metallic particles by KUBO [2.17], on complementary aspects of the NMR and CESR experiments by KNIGHT [2.18], and on phonons in finite crystals by BALTES [2.3].

Moreover, the structure and dynamics as well as the preparation of simple microclusters is reviewed by HOARE [2.19]. MUELLER-KRUMBHAAR [2.20] has studied the simulation of small systems in a recent book *Monte Carlo Methods in Statistical Physics*. As for the statistical theory of nucleation, condensation, and coagulation, we refer to an article by BINDER and STAUFFER [2.21]. The thermal radiation laws for finite blackbody cavities are reviewed by BALTES [2.22,23]. There are several comprehensive articles on the surface dynamics of dielectrics and the optical phonons in finite ionic crystals, such as by KÄLIN and KNEUBÜHL [2.24], KLIEWER and FUCHS [2.25], GENZEL [2.26], and RUPPIN and ENGLMAN [2.27]. We finally mention the reviews by HUFFMAN [2.29] and WICKRAMASINGHE and NANDY [2.28] on the physics of interstellar dust.

2.2 Perfect Gases in Finite Boxes of Regular Shape

The simplest model of a microparticle is that of a small box enclosing a perfect quantum gas, i.e., an ensemble of non-interacting particles (such as electrons) or elementary excitations (such as phonons). The study of the statistical physics of such a system leads to *Weyl's problem*, i.e., the question of finding the distribution of the eigenvalues of the pertinent wave equation in a finite domain of given shape and with given boundary conditions. In Sect.2.2.1 Weyl's problem is briefly reviewed. Sections 2.2.2-5 are devoted to applications of Weyl's problem, namely the quantum-size effect of the vibrational specific heat, the radiation laws for small blackbody cavities, the electronic magnetic moment in the quantum-size regime, the thermodynamic relations in the quasi-continuum regime, and the Bose-Einstein condensation.

2.2.1 Weyl's Problem

By Weyl's problem we understand the problem of calculation the eigenvalue distribution for the wave equation [2.14]. Weyl considered three different types of boundary-value problems.

a) The *scalar* problem, $(\nabla^2 + k^2)U = 0$, related to the perfect quantum gas of particles of mass m by virtue of $k = (2m\varepsilon)^{\frac{1}{2}}/\hbar$, and to the acoustic resonator by virtue of $k = \omega/v$. Here, k denotes the wavenumber, U the wave amplitude, ε the energy, ω the circular frequency, and v the sound velocity.

b) The *electromagnetic* problem, $(\nabla^2 + k^2)\underline{E} = 0$, $\mathrm{div}\,\underline{E} = 0$, $k = \omega/c$, with $\underline{E}$ denoting the electric field and c the speed of light. This problem is basic to quantum electrodynamics in a finite box.

c) The *elastic* problem,

$$[(v_\ell^2-v_t^2)\text{grad div}+v_t^2\nabla^2+\omega^2]\ \underline{U} = 0 \quad , \tag{2.1}$$

relevant to acoustical phonons in the elastic continuum model. Weyl's nonphysical boundary condition div $\underline{U}$ = 0 allows the separation into transverse modes with velocity v_t and longitudinal modes with velocity v_ℓ irrespective of the size of the domain. *Clamped* boundaries with $\underline{U}$ = 0 (an approximation for soft, high-density microparticles in contact with hard, low-density matrix material) as well as *stress-free* boundaries (corresponding to microparticles with free surface) with the true physical boundary condition

$$\sum_\ell \delta_{k\ell}(v_\ell^2-v_t^2)\ \text{div}\ \underline{U}+v_t^2(\partial_\ell U_k+\partial_k U_\ell) = 0 \tag{2.2}$$

lead to the scrambling of t and ℓ modes. The boundary condition (2.2) also leads to Rayleigh modes that are localized near the surface. Similar phenomena occur in the intermediate case of *contact* boundaries. It is only in the bulk limit that the boundary conditions are unimportant and that t and ℓ modes are well defined.

We recall that *Weyl's theorem* asserts that the bulk density of states is independent of the shape of the domain; the model density $D(\omega)$ shows the asymptotic behavior $D(\omega) \sim \text{const}\ V\omega^2$ as $V \to \infty$ or $\omega \to \infty$. It is well known that this result is the basis of the bulk statistical physics of perfect gases and leads to Debye's T^3 law and Planck's radiation law. In view of the open problems of microparticle physics, much effort has been spent on establishing asymptotic expansions for the mode density, where Weyl's theorem provides the bulk term and where higher-order terms account for the surface and shape effects. Since the substantial progress of Weyl's problem is already covered in the book by BALTES and HILF [2.14], only a few typical or very recent results and references are mentioned here. The eigenvalue distribution is completely known for the scalar [2.30-38] and the electromagnetic [2.11,12,22,23,34,35,37,39-43] problem, but only the first-order correction or surface term is known for the elastic problem [2.44-48] (Sect.2.2.2).

As an illustrative example we present here the scalar problem for the *cube-shaped domain* (edge length L) with Dirichlet (U = 0) or Neumann ($\partial U/\partial n = 0$) boundary conditions. In terms of the wavenumber k, the mode density D(k) can be defined by

$$\begin{aligned} D(k) = {} & \frac{1}{8}\sum_{n_1,n_2,n_3=-\infty}^{+\infty} \delta[k-\pi(n_1^2+n_2^2+n_3^2)^{1/2}/L] \\ & \pm \frac{3}{8}\sum_{n_1,n_2=-\infty}^{+\infty} \delta[k-\pi(n_1^2+n_2^2)^{1/2}/L] \\ & + \frac{3}{8}\sum_{n_1=-\infty}^{+\infty} \delta(k-\pi|n_1|/L) \pm \frac{1}{8}\,\delta(k) \quad . \end{aligned} \tag{2.3}$$

This mode density can be rewritten in terms of a complete expansion, viz.,

$$D(k) = L^3k^2/2\pi^2 \pm 3L^2k/4\pi + 3L/4\pi \pm \frac{1}{8}\,\delta(k)$$

$$+ (L^3k^2/2\pi^2) \sum_{n_1,n_2,n_3}' \mathrm{sinc}[2kL(n_1^2+n_2^2+n_3^2)^{\frac{1}{2}}] \pm (3L^2k/4\pi) \sum_{n_1,n_2}' J_0[2kL(n_1^2+n_2^2)^{\frac{1}{2}}]$$

$$+ (3L/4\pi) \sum_{n_1}' \cos(2kLn_1) \quad , \qquad (2.4)$$

where $\mathrm{sinc}\, x \equiv x^{-1}\sin x$ and J_0 denotes the Bessel function of order zero, and where the primes indicate that the terms corresponding to $n_i = 0$ for all i are omitted in the summation. The positive signs apply in the case of the free surface (Neumann) and the negative signs in the case of the clamped surface (Dirichlet).

In physical applications, an expression of the type (2.4) is usually subject to some averaging procedure such as the convolution with a spectral window of finite width or another integral transform with appropriate kernel. In the *quasi-continuum* regime, where the distance between adjacent eigenvalues is small compared with the spectral resolution, the strongly oscillating sums in (2.4) are smoothed out by such averaging procedures and hence can be dispensed with. In this regime, the first four terms in (2.4) are sufficient. In general, they can be identified as the volume, the surface, the edge (and curvature), and the corner (and connectivity) terms, respectively. In the *quantum-size* regime, however, the oscillatory terms in (2.4) are relevant, if not dominant, and it may be easier to use directly the original expression of the type (2.3).

2.2.2 Vibrational Specific Heat

Debye's T^3 law describing the vibrational contribution to the low-temperature specific heat of solids seems to be the first classical relation of bulk solid-state physics that was called in question by a quantum-size effect. As early as 1921, PLANCK [2.49] and SCHAEFER [2.50] predicted that the T^3 law would break down at least for temperatures so low and crystallites so small that only the fundamental mode of frequency ω_1 is appreciably excited as compared to the modes with higher frequency. In this *single-mode regime*, the vibrational specific heat would be proportional to $T^2\exp(-\hbar\,\omega_1/k_BT)$. This type of specific heat may be observed if $T \ll \hbar\,\omega_1/k_B$ or for microparticles with diameter $d \ll 100$ Å/T[K]. In the *quantum-size regime*, $d \approx 100$ Å/T[K], the specific heat is obtained [2.4,5] by evaluating an exponential sum of the type

$$(d/dT) \sum_{i=1}^{M} \hbar\omega_i\, [\exp(\hbar\omega_i/k_BT)-1]^{-1} \qquad (2.5)$$

with M denoting the number of vibrational degrees of freedom. In the *quasi-continuum regime*, $d \gg 100$ Å/T[K], the specific heat may be obtained by integration of an appropriate mode density, whose leading term yields Debye's T^3 expression. As compared to the bulk value, a *decrease* of the specific heat is predicted in the single-mode regime, whereas an *enhancement* is expected for free particles in the quantum-size and quasi-continuum regimes.

In principle, the resonance frequencies ω_i can be obtained from an appropriate lattice model or from the elastic continuum model. Pursuing the latter model, one has to apply Weyl's third problem [case (c) in Sect.2.2.1]. Unfortunately, no complete expansion of the type (2.4) is known in this case. The best result available [2.44-48] for the stress free boundary obeying (2.2) is the asymptotic expansion

$$D(\omega) \sim \frac{V\omega^2}{2\pi^2}\left(\frac{2}{v_t^3}+\frac{1}{v_\ell^3}\right)+\frac{A\omega}{8\pi}\,\frac{2v_t^4-3v_t^2v_\ell^2+3v_\ell^4}{v_t^2v_\ell^2(v_\ell^2-v_t^2)} \tag{2.6}$$

with A denoting the surface area of the microparticle. The first term is the bulk mode density, which includes the two transverse branches and the longitudinal branch. The second term describes the surface correction, and its complicated dependence on the bulk parameters v_t and v_ℓ reflects the mode scrambling. The next term (which presumably would involve edge lengths and curvatures) is not known. The corresponding quasi-continuum specific heat reads

$$C = C_{\text{Debye}} + \text{const.}AT^2 + O(T) \quad . \tag{2.7}$$

From these results, one would expect that microparticles with free surfaces in the quasi-continuum regime show an *enhancement* of the acoustical mode density and low-temperature specific heat together with the corresponding *decrease of the Debye temperature*. This is indeed observed in neutron scattering and calorimetric experiments (e.g.,[2.51] and references therein).

Calculations of the vibrational specific heat in the quantum-size regime [2.5, 52,53] were stimulated by measurements of NOVOTNY et al. [2.7,54] and later confirmed by the experiments (Fig.2.2) of COMSA et al. [2.4]. In the model proposed by BALTES and HILF [2.5], the microparticles are treated as homogeneous elastically vibrating spheres (radius R) with free surfaces. In view of the difficulty of the true elastic problem, the scalar version of Weyl's problem [case (a) in Sect. 2.2.1] with the Neumann boundary condition is applied instead, and the *microparticle* or *effective sound velocity* $v = v(R,T)$ is considered as an adjustable parameter. The resulting vibration specific heat reads [2.5] (eigenfrequencies $\omega_{\ell,s} = v\, a'_{\ell,s}/R$)

$$C(T,R) \propto \sum_\ell \sum_s \frac{3(2\ell+1)k_B\xi^2\exp\xi}{4\pi R^2[\exp(\xi-1)]^2} \tag{2.8}$$

with $\xi = \hbar v a'_{\ell,s}/k_B RT$, where $a'_{\ell,s}$ denotes the sth zero of the derivative of the spherical Bessel function of order ℓ. The largest zero a'_{max} to be taken into account is determined by the number of atoms contained in the microparticle. This leads to the *microparticle* or *effective Debye temperature* $\Theta(R,T) = \hbar a'_{max} v(R,T)/k_B R$. Quantitative agreement between the experimental data [2.4,53,54] and (2.8) is obtained on the basis of a size and temperature dependence of the sound velocity, or the Debye temperature. The resulting $\Theta(R,T)$ and $v(R,T)$ are smaller than the corresponding bulk values and are supported by data from other experiments, such as X-ray diffraction [2.55] and measurement of the superconducting transition temperature [2.56]. Averaging of (2.8) over the appropriate size distribution does not lead to appreciably different values of Θ and v [2.4]. Further interesting studies of acoustical vibrations of microparticles use electron diffraction (e.g.,[2.57]) and Mössbauer spectroscopy (e.g.,[2.58]).

2.2.3 Radiation Laws

BIJL [2.59] seems to have been the first to discuss the case of thermal radiation enclosed in a microcavity in the same way as PLANCK [2.49] and SCHAEFER [2.50] had studied the single-mode regime of the vibrational specific heat of microparticles. He noticed that the classical radiation laws of Kirchhoff, Stefan and Boltzmann, Wien, and Planck no longer apply at temperature T so low and cavity volumes V so small that $TV^{1/3} \lesssim \hbar c/k_B \approx 0.23$ K cm. Deviations from the classical radiation laws must be expected unless $TV^{1/3}$ is very large compared with $\hbar c/k_B$. In the last decade, thermal radiation in small cavities was studied in much detail by CASE and CHIU [2.60], BALTES and coworkers [2.11,12,22,23,35,37,39-41,61-64], and BALIAN and BLOCH [2.42,43] and ECKHARDT [2.65-68]. Most of this work is concerned with empty cavities showing ideal (i.e., lossless) walls. Only a few papers are devoted to the more involved problem of a small cavity filled with dielectric material [2.60,66,68] or with walls of finite conductivity [2.65,67]. In this section, only a few typical results for the ideal-wall case can be reported.

For simplicity, we consider the example of the *cube-shaped empty lossless cavity* of edge length L. The complete expansion of the "electromagnetic" mode density reads [2.35,37,40]

$$D(k) = \pi^{-2}L^3k^2 - 3L/2\pi + \frac{1}{2}\,\delta(k) + \pi^{-2}L^3k^2 \sum_{n_1,n_2,n_3=-\infty}^{+\infty} \mathrm{sinc}[2kL(n_1^2+n_2^2+n_3^2)^{\frac{1}{2}}]$$
$$+ (3L/2\pi) \sum_{n_1=-\infty}^{+\infty} \cos(2kLn_1) \quad , \tag{2.9}$$

a result analogous to the "scalar" mode density (2.4). The first term in (2.9) is the bulk term leading to the classical radiation laws. Contrary to the scalar and the elastic case, the surface term of the electromagnetic mode density of an empty lossless cavity vanishes [2.39,42,62]. The corresponding total thermal radiation energy E in the quasi-continuum regime is given by [2.40,60,63]

$$(L/\hbar c)\ E(\eta) \sim (\pi^2/15)\eta^4 - (\pi/4)\eta^2 + \eta/2 + \dots \qquad (2.10)$$

with $\eta \equiv k_B TL/\hbar c$. The above three terms are an excellent approximation for $TL \gtrsim 1$ K cm, as is shown in [2.63]. For $TL \lesssim 1$ K cm, the oscillatory or quantum-size terms of (2.9) are indispensable. The resulting full total radiation energy is shown in Fig.2.6. The full analytical formula for $E(\eta)$ can be found in [2.12,22].

In a microcavity, size and shape effects are expected for all the quantum-electrodynamical phenomena that involve the electromagnetic mode density. Examples are the field correlations [2.12,40,41,64,66,67] (Fig.2.7), spontaneous emission [since the Einstein coefficient A is proportional to $\omega\, D(\omega)$], fluorescence lifetime [2.69], the Casimir effect (e.g.,[2.70]), and the Lamb shift [2.71].

Application of the above microcavity results to freely radiating particles in an aerosol particle is not straightforward. In general, the "background" radiation "seen" by the microparticles plays an important role. In principle, size effects may be expected at room temperature for particles whose diameter is of the order of 10 μm or less. This estimate is in agreement with current studies of the thermal FIR emission of dielectric microcrystals due to optical phonons [2.24] (Sect.2.3). Finally it may be mentioned that the radiation of partially coherent sources of finite size is presently investigated [2.72] in the entirely different framework of diffraction and coherence theory. One interesting result is that only the size (compared to the wavelength of the radiation) but also the state of the coherence (correlation length) of the source is crucial to the resulting spatial radiation pattern.

2.2.4 Electronic Magnetic Moments

Although electronic properties of metallic microparticles are further considered in Sects.2.4-6, we present in this section the case of the magnetic moment of indium particles in the quantum-size regime measured and interpreted by MEIER and WYDER [2.6,73], since the experimental results (Fig.2.3) can be accounted for in terms of the orbital magnetism of free electrons in a spherical square-well potential of infinite depth, i.e., by applying Weyl's scalar problem with Dirichlet boundary conditions. The energy eigenvalues are determined by the zeros $a_{n,\ell}$ of the spherical Bessel function of order ℓ, viz., $\varepsilon_{n,\ell} = \hbar^2 a_{n,\ell}^2/2mR^2$, where R denotes the particle radius and ℓ determines the related orbital momentum. By properly accounting for the highest occupied level, ε_{n_0,ℓ_0} (which is the only magnetically active level) and the pertinent level spacing and degeneracy, the magnetization is calculated from the grand canonical partition function. The chemical potential ξ_N is determined from the equation

$$N = \sum_{m=-\ell}^{+\ell} \sum_{s=-\frac{1}{2}}^{+\frac{1}{2}} \left(\exp\left\{\left[(m+2s)\mu_B H - \xi_N(H,T)\right]/k_B T\right\} + 1\right)^{-1} \quad , \tag{2.11}$$

where N denotes the number of electrons occupying the level ε_{n_0,ℓ_0} and H the magnetic field; nonzero magnetic moment is obtained for $1 \le N \le 2(2\ell_0 + 1)$. Averaging over the possible N, one finds the magnetization

$$M(H,T) = \frac{\mu_B}{2(2\ell_0+1)} \sum_N \sum_m \sum_s \frac{m+2s}{1+\exp\{\mu_B H[m+2s-\bar{\xi}_N(H,T)]/k_B T\}} \tag{2.12}$$

with $\bar{\xi}_N = \xi_N/\mu_B H$. This model calculation is able to describe the rapid increase of the magnetization as a function of H as shown in Fig.2.3. Exact agreement with the experiment is achieved by considering slightly nonspherical particles, but the exact shape is not vital for explaining the gross features of the measured magnetization. As was checked by MEIER [2.73], theories based on spin paramagnetism and using either equally spaced levels or random (Poisson) distributions of nondegenerate levels (Sect.2.4) cannot account for the experimental results. Of course, the electronic contribution to the specific heat can also be calculated in the framework of the above model [2.74].

2.2.5 Thermodynamic Relations and Bose-Einstein-Condensation

From the above results it is easily inferred that the classical thermodynamic relations are not valid for microparticles in contact with some reservoir - neither in the quasi-continuum nor in the quantum-size regime. For example, the well-known relations $F = -E/3$, $S = 4E/3T$, $C_V = 4E/T$ for the *thermal radiation field* in a cavity (with F denoting the free energy, S the entropy, C_V the specific heat at constant volume) apply only in the bulk limit (very large $TV^{1/3}$). For small cavities, only the relation $PV = E/3$ (with P denoting the radiation pressure) remains valid [2.39], at least in the quasi-continuum regime.

HILF [2.75] studied the thermodynamics of *nonrelativistic noninteracting particles* of mass m, bounded by perfectly reflecting rigid walls, in the quasi-continuum regime (see also [Ref.2.14,Sect.VII.2]). In this case, Weyl's scalar problem with Dirichlet boundary conditions applies. The pertinent density of states is derived from results of the type (2.4) by inserting $k = (2m\varepsilon)^{\frac{1}{2}}/\hbar$, where ε denotes the variable corresponding to the energy eigenvalues of the Schrödinger equation. Omitting the spin degeneracy, the density of states is given by

$$\frac{\hbar}{2m} D(\varepsilon) \sim \frac{V}{4\pi^2}\left(\frac{2m\varepsilon}{\hbar^2}\right)^{\frac{1}{2}} - \frac{A}{16} + \frac{\Lambda}{16\pi^2}\left(\frac{2m\varepsilon}{\hbar^2}\right)^{-\frac{1}{2}} + \dots \quad , \tag{2.13}$$

where A denotes the surface area of the enclosure and the length Λ characterizes the edges and curvatures of the box (e.g., $\Lambda = 6\pi L$ for a cube of edge length L and $\Lambda = 8\pi R/3$ for a sphere of radius R). It is useful to introduce dimensionless thermodynamic quantities such as $\tilde{E}$, $\tilde{F}$, $\tilde{T}$, $\tilde{\mu}$ by virtue of the definitions

$$\tilde{E}/E = 2mV^{2/3}/\hbar, \qquad \tilde{T}/T = 2mV^{2/3}/\hbar,$$

$$\tilde{F}/F = 2mV^{2/3}/\hbar, \qquad \tilde{\mu}/\mu = 2mV^{2/3}/\hbar, \tag{2.14}$$

where μ denotes the chemical potential, but to keep the particle number N and the entropy S. From the density of state (2.13), the function N, $\tilde{E}$, and $\tilde{F}$ are calculated in terms of the Planck-Fermi integrals

$$J_n(x) = \int_0^\infty dy\, y^n\left[x^{-1}\exp(x^2)-j\right]^{-1} \tag{2.15}$$

with $x = \exp(\mu/T) \equiv \exp(\tilde{\mu}/\tilde{T})$ and $j = 1$ (Bose-Einstein statistics), $j = -1$ (Fermi-Dirac statistics) or $j = 0$ (Maxwell-Boltzmann statistics), e.g.,

$$\tilde{E}(x,\tilde{T}) \sim (2\pi^2)^{-1}\,\tilde{T}^{5/2}J_4(x) - (8\pi)^{-1}AV^{-2/3}\tilde{T}^2J_3(x) + (8\pi)^{-1}\Lambda V^{-1/3}\tilde{T}^{3/2}J_2(x) + \ldots \tag{2.16}$$

Using (2.14), from (2.16) and the analogous results for $\tilde{N}(x,\tilde{T})$ and $\tilde{F}(x,\tilde{T})$, complete thermodynamic information is obtained, in principle. One finds that $PV = 2E/3$ is the only "bulk" relation for the nonrelativistic quantum gas that remains valid for the finite system characterized by (2.13).

Near the boundary walls, the particle density and the energy density on the position, and hence *surface tension* and *shape tension*, $\sigma = \partial F/\partial A$ and $\kappa = \partial F/\partial \Lambda$, exist. In the quasi-continuum regime, one finds a microparticle relation, which is analogous to the Gibbs-Duhem relation $G = \mu N$ of the "bulk" gas, viz.,

$$G = \mu N + \tfrac{1}{3}\sigma A + \tfrac{2}{3}\kappa\Lambda \quad , \tag{2.17}$$

where G denotes the Gibbs function. Thus the gas in the finite box can be called "locally homogeneous", since the *differential* bulk relation $\partial G/\partial N = \mu$ holds, but by virtue of (2.17) it is not "globally homogeneous". However, as HILF [2.75] shows, the quantities $\tilde{A} = A^3$ and $\Lambda = \Lambda^{3/2}$ play the role of extensive surface and shape variables that can be used in order to generalize the set of state variables so that the generalized system is globally homogeneous. In this sense, a generalized thermodynamics can be established in the quasi-continuum regime. In the quantum-size regime, however, it is not possible to establish a thermodynamic description independent of the type of statistics [2.75].

For the *number dependence* of small-crystal thermodynamic properties we refer to a paper by ABRAHAM and KORTZEBORN [2.76]. A study of the *surface contribution to the pressure* of free quantum gases was published by NENCIU [2.77]. The fluctuations of the temperature in small system is discussed by McFEE [2.78].

GREENSPOON and PATHRIA [2.79], and CHABA and PATHRIA [2.80] study the *Bose-Einstein condensation* in cuboid-shaped boxes both in the quasi-continuum [2.79,80] and in the quantum-size regime [2.80], where the thermal wavelength of the particles is of the order of the edge lengths L_i of the enclosure (see also [2.36]). Finite-size corrections play an important role in the critical region. Instead of the discontinuities ("sharp" extrema) existing only in the bulk limit, a smooth maximum of C_V/N and a smooth minimum of $\partial^2\mu/\partial T^2$ are found. GREENSPOON and PATHRIA [2.79] found a *law of corresponding states,* according to which two Bose-Einstein systems similar in shape, but different in size, are in corresponding states when their reduced chemical potentials $\tilde{\mu}$ [as defined by HILF's transformation (2.14)] are equal. Reference [2.80] discusses the Bose-Einstein condensation at *constant pressure* P. The critical temperature T_c differs from the bulk limit $T_c(\infty)$ and turns out to be a complicated function of the total number N of bosons. At temperatures somewhat below $T_c(\infty)$, the quantum-size effect is dominant and the volume of the system, together with the specific heat at constant pressure, C_P, becomes proportional to $N^{5/3}$.

The *statistical physics* of general finite systems is beyond the scope of this chapter, and we refer to the original literature. We mention the critical temperature of finite systems [2.81], the finite Ising models [2.82], finite systems of weakly coupled chains [2.83], size effects on the measurement of transport quantities [2.84] and on the momentum autocorrelation function [2.85], and Brownian motion and the velocity autocorrelation [2.13,86,87].

2.3 Optical Phonons in Dielectric Microparticles

Size and shape effects are discussed with emphasis on the FIR emission of dielectric microparticles. In Sect.2.3.1 the vibrational dispersion scheme is briefly discussed. Next, the theory of optical modes in microparticles is summarized (Sect.2.3.2). Finally, Sect.2.3.3 is devoted to typical experimental results. Since excellent reviews [2.24-27] are available, only a condensed overview is presented here.

2.3.1 Dispersion Scheme and Gap Modes

We recall that the vibrational modes of solids are described in terms of the frequency ω and the wave vector $\underline{k}$. The set of possible values of ω and $\underline{k}$ is organized into a dispersion scheme showing a limited number of branches, $\omega_i(\underline{k})$, with i denoting the branch index, as a function of the wave number $\underline{k}$, which is a quasi-continuous variable in the bulk case. In the dispersion scheme of biatomic bulk crystals, one can distinguish *acoustical* branches (indices: TA,LA) whose frequencies go to zero as $\underline{k}$ goes to zero, and *optical* branches (indices: TO,LO) which exhibit nonzero frequencies $\omega_{TO} = \omega_{TO}$ ($\underline{k}$ = 0) and $\omega_{LO} = \omega_{LO}$ ($\underline{k}$ = 0). The parameters ω_{TO} and ω_{LO} are crucial for the spectral properties of the bulk crystal in the infrared spectral region. The optical branches are separated from the acoustical branches by a gap of vanishing frequency density $D(\omega)$, the *acousto-optical gap*.

As discussed in Sect.2.2.2, the distinction between transverse (TA) and longitudinal (LA) acoustic modes of low frequency is meaningless in microcrystals because of the *mode scrambling* induced by a stress-free boundary. Besides the uniform modes, a different type of acoustical vibration, the Rayleigh modes (which are localized near the surface) appear. Moreover microcrystals show a rich variety of optical branches, i.e., of frequencies $\omega_i(\underline{k} = 0) > 0$, as described in Sect.2.3.2. In particular, microparticles exhibit a number of optical *gap modes*, i.e., vibrational modes whose frequencies bridge the acousto-optical gap existing in the bulk crystal. The gap modes manifest themselves not only in the infrared spectrum (Fig.2.1) but also in *neutron scattering* experiments. For example, RIEDER [2.88] and RIEDER and DREXEL [2.89] observed gap modes in the inelastic neutron scattering spectra of MgO (average diameter: 160 Å) and TiN (average diameter: 300 Å) microparticles. The resulting frequency density $D(\omega)$ of the TiN microparticles is shown in Fig.2.8. Besides the size effects on the optical frequency density, Fig.2.8 also shows an enhancement of the low-frequency acoustical mode density, which corresponds to the specific-heat enhancement studied in Sect.2.2.2.

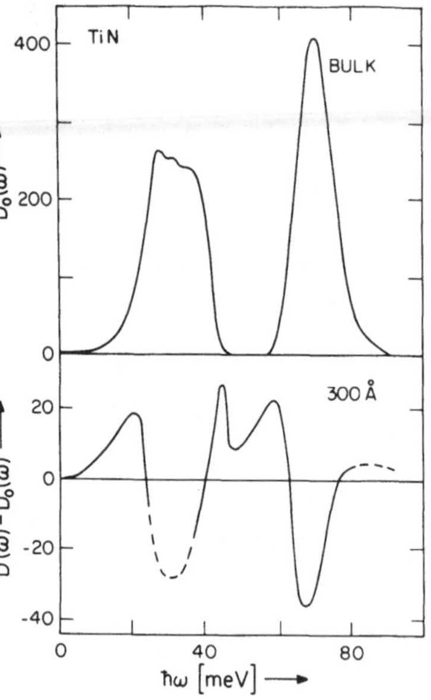

Fig. 2.8. Phonon frequency density $D_0(\omega)$ of bulk TiN (above) and difference $D(\omega) - D_0(\omega)$ between 300 Å TiN microparticle and bulk densities (below). (Adapted from [2.89])

2.3.2 Summary of Theoretical Results

FRÖHLICH [2.90] was the first to predict an optical gap mode in dielectrics, namely a uniform optical vibrational mode of the frequency

$$\omega_F = [(\omega_{TO}^2 + \omega_{LO}^2)/2]^{\frac{1}{2}} \quad , \tag{2.18}$$

for small spheres (radius $R \ll \lambda_{TO} = 2\pi c/\omega_{TO}$) of diatomic ionic crystal material. ENGLMAN and RUPPIN [2.91] developed the full theory of long-wave optical phonons in small spheres in the limit $R \ll \lambda_{TO}$. Besides the eigenfrequencies ω_{TO} and ω_{LO} (as in the bulk case), they find a set of modes with frequencies

$$\omega_{S,\ell} = \{[(\ell+1)\omega_{TO}^2 + \ell\omega_{LO}^2]/2\ell+1\}^{\frac{1}{2}} \quad . \tag{2.19}$$

The case $\ell = 1$ corresponds to a uniform mode of the type predicted by FRÖHLICH [2.90], whereas the cases $\ell \geq 2$ correspond to modes that are localized near the surface. We notice that *Fröhlich's* frequency (2.18) is obtained from (2.19) in the limit $\ell \to \infty$.

The general theory for diatomic crystalline spheres of arbitrary size, i.e., radius R not necessarily much smaller than the wavelength λ_{TO}, has to take into account the retardation of the Coulomb forces. This was achieved by FUCHS and KLIEWER [2.92] and leads to the following spectrum (for small wavenumbers $\underline{k}$):

a) the longitudinal mode with $\omega = \omega_{LO}$ as in bulk material;

b) the "high-frequency" transverse modes, i.e., a family of modes with frequencies $\omega_i > \omega_{LO}$. They show the limit behavior $\omega_i \to \infty$ for $R \ll \lambda_{TO}$ and $\omega_i \to \omega_{LO}$ for $R \geq \lambda_{TO}$;

c) The "low-frequency" transverse modes of frequencies $\omega_i < \omega_{TO}$ with $\omega_i \to \omega_{TO}$ for $R \ll \lambda_{TO}$ and $\omega_i \to 0$ for $R \gg \lambda_{TO}$;

d) The "intermediate" transverse modes with frequencies $\omega_i \to \omega_{S1}$ as given by (2.19) in the case $R \ll \lambda_{TO}$ and $\omega_i \to 0$ for $R \gg \lambda_{TO}$.

FUCHS and KLIEWER [2.92] also calculate the infrared absorption cross section of spheres of diatomic ionic crystal material. They predict that only the mode

$$\omega_{S1} = (2\omega_{TO}^2/3 + \omega_{LO}^2/3)^{\frac{1}{2}} \tag{2.20}$$

can actually be observed as an absorption (or emission) peak in the case $R \ll \lambda_{TO}$, whereas the many other modes do not absorb appreciably.

FUCHS [2.93] developed a theory for the complex electric susceptibility of microparticles of average diameter much smaller than λ_{TO} but of arbitrary shape. He thus is able to account for microcrystal shapes that occur in many experiments, such as cubes or prisms. Another case close to the experimental situation, namely that of spherical particles randomly distributed in a nonabsorbing medium, was studied by GENZEL and MARTIN [2.94] under the assumption that the spacing between the particles is much larger than the particle diameter, and that again $R \ll \lambda_{TO}$. Combining these two theories KÄLIN and KNEUBÜHL [2.24] calculated the emissivity of a layer of KCl cubes and spheres. For the spherical particles they find one strong resonance similar to that predicted by FUCHS [2.93] for a single sphere. For the cubes, however, they obtain several side resonances besides the main peak (which is similar to that of the spheres). The side resonances correspond to shape-dependent modes, which are absent in the case of sphere because of its high symmetry.

2.3.3 Far-Infrared Measurements

Most of the FIR studies of dielectric microparticles are made by *absorption* spectroscopy. As examples we mention the FIR absorption measurements of alkali halide microcrystals by BRYKSIN et al. [2.95], NOVÁK [2.96], NAHUM and RUPPIN [2.97], CLIPPE et al. [2.98] and MARTIN [2.99], and the absorption spectra of amorphous quartz microspheres measured by STEYER et al. [2.100].

A less common, but interesting method is the measurement of the spectral *thermal emission* of microcrystals introduced by KÄLIN et al. [2.101] in 1970. Later, KÄLIN and KNEUBÜHL [2.24] made a systematic investigation of the FIR emission of KCl microcrystals on a (poorly emitting) metallic substrate, with various microcrystal sizes (2 μm and 10 μm average diameter) and shapes (e.g., cubes and octahedra) and different covering rates. The main features of the thermal emission spectra are as follows (Fig.2.1).

a) A strong resonance is always found near $\omega = \omega_{TO}$. Such a mode is expected from theory for larger particles (e.g., 10 μm diameter), but not for the 2-μm particles. In the latter case, it can possibly be ascribed to clustering effects.

b) Another strong resonance occurs in the gap between ω_{TO} and ω_{LO}. It can be interpreted as the fundamental "intermediate" mode analogous to the mode ω_{S1} of the microsphere, see [2.20].

c) Weaker resonances in the gap between ω_{TO} and ω_{LO} can be explained in terms of the shape-dependent surface modes discussed in Sect.2.3.2.

Finally, we refer to a few *Raman* studies of microparticles due to SUN and ANDERSON [2.102], ETZ et al. [2.103], and KERKER et al.[2.104].

2.4 Electronic Heat Capacity and Magnetic Susceptibility of Metallic Microparticles

Thermodynamic properties, such as heat capacity and spin paramagnetism, provide a useful testing ground for the study of the electronic excitation spectrum of solids. In this section we discuss these properties for the case of *metallic microparticles*. First, the early theoretical ideas are reviewed in Sect.2.4.1. Next, in Sects.2.4.2-4 we present an elementary derivation of the temperature dependence of the heat capacity and the spin paramagnetism. In Sect.2.4.5 we introduce the concepts and methods of ensemble averaging over random energy levels. The comparison of predicted thermodynamic properties is discussed in Sect.2.4.6,7. Finally, the effects of spin-orbit coupling on the spin susceptibility of metallic microparticles with an even number of electrons is discussed in Sect.2.4.8, together with the effects of level quantization on spin relaxation processes. An extension of these concepts to the Kondo effect in metallic microparticles is discussed as well.

2.4.1 Historical Background

In a bulk metal the electron energy spectrum is continuous and Fermi-Dirac statistics is appropriate for the conduction electrons. The well-known consequence of this is the electron heat capacity, which is proportional to the absolute temperature T, and the Pauli paramagnetic susceptibility, which is independent of T [2.105]. In a metallic microparticle (m.m.) the finiteness of the size brings about some new effects that modify the above thermodynamic properties. The most important one is the quantization of the states of the valence electrons due to the surface boundary conditions, which leads to the discreteness of the electron energy levels. As a result of this, a finite energy gap opens between the electronic ground state and the first excited state that is reminiscent of the energy spectrum of dielectric crystals. For the latter the electronic contributions to the thermodynamic properties are exponentially small, being proportional to $T^{-2}\exp(-\Delta/k_BT)$, where Δ is the energy gap [2.106]. FRÖHLICH [2.107] was the first to discuss this thermodynamic anomaly for the m.m. Assuming particles with *equidistant electron levels*, he predicted that the electronic heat capacity should decrease exponentially at very low temperatures in contrast to the linear law of bulk metals. The related anomaly of the low-temperature spin susceptibility in m.m. was pointed out later by GREENWOOD et al. [2.108]. Subsequently KUBO [2.109] came independently to the same conclusion regarding the spin susceptibility. Moreover, he introduced two important concepts, which eventually brought the theory of the m.m. to its present state.

1) The first concept is based on the observation that in practice, one measures the thermodynamic properties of a collection of a large number of m.m. of varying size and shape. Even in an ideal sample composed of particles of the same size, the irregular shapes are responsible for variations of energy-level spectra from particle to particle. KUBO [2.109] introduced a statistical description of the spectrum based on a distribution function $P(\Delta)$ defined as

$$P(\Delta)d\Delta = Pr(\Delta,\Delta+d\Delta) \quad , \qquad (2.21)$$

where $Pr(\Delta,\Delta+d\Delta)$ is the probability to find the successive level spacing Δ in the interval $(\Delta,\Delta+d\Delta)$. The average level spacing near the highest occupied level is assumed equal to the inverse bulk density of states

$$\delta = \int \Delta P(\Delta)d\Delta \simeq 4\, E_F/3N \quad , \tag{2.22}$$

where E_F is the Fermi level of the bulk metal and N is the number of electrons in the particles. Equation (2.22) implies that δ varies as R^{-3}, where R is the radius of the m.m. For example, an aluminum particle with radius R = 25 Å contains about 1.2×10^4 electrons so that $\delta \simeq 1.3 \times 10^{-4}$ eV. A more detailed discussion of the distribution function $P(\Delta)$ is presented in Sect.2.4.5. We would like to point out, however, that the popular expression (2.22) is a bulk-type result that does not do full justice to the microparticle situation. From the Dirichlet-problem mode density D(k) of the type (2.3), one may infer that in m.m. δ goes as $N^{-1/3}$ or R^{-1} rather than as N^{-1} or R^{-3} and hence is much smaller than predicted by (2.22). This could explain the absence of quantum-size effects in certain microparticle experiments [2.167,172].

2) The second concept introduced by KUBO [2.109] is that of electrical neutrality. The energy required to put an extra electron on an isolated m.m. of radius R is of the order of

$$U \simeq e^2/2R \quad . \tag{2.23}$$

For R = 25 Å this amounts to U ≃ 0.26 eV, which is three orders of magnitude above the thermal energy k_BT at T = 1 K. Consequently, each of the m.m. remains neutral at such low temperatures. The basic statistical implication of this is that the electronic partition function of a m.m. must be calculated with the use of the canonical, rather than the grand canonical ensemble. This also implies that one cannot apply the familiar Fermi distribution functions to calculate the thermodynamic properties. In other words, m.m. are assumed to be in contact with a thermal, but not an electron, reservoir: while energy can be exchanged, electrons cannot.

The derivation of the canonical partition function as a projection from the grand canonical one by means of contour integration was presented by KUBO [2.109], but its discussion is beyond the scope of this chapter. Instead we present below a simple derivation of the thermodynamic properties which at very low temperatures (i.e., $k_BT \ll \delta$) yields the temperature dependences in agreement with a more sophisticated calculation by DENTON et al. [2.110]. There is an important distinction, both experimentally and theoretically, between the magnetic susceptibility of a particle with an even number of electrons and that of an "odd" particle [2.109]. Hence it is useful to consider the thermodynamics of the "even" and "odd" particles separately. We also note that this exposition is confined to the m.m. in a normal state. The superconducting m.m. are the subject of Sect.2.6.

2.4.2 Free Energy of an "Even" Particle

Following KUBO [2.17] we regard the m.m. as a giant molecule with the one-electron energy levels...$\varepsilon_{-2},\varepsilon_{-1},\varepsilon_0,\varepsilon_1,\varepsilon_2$... . In a particle with an even number of electrons,

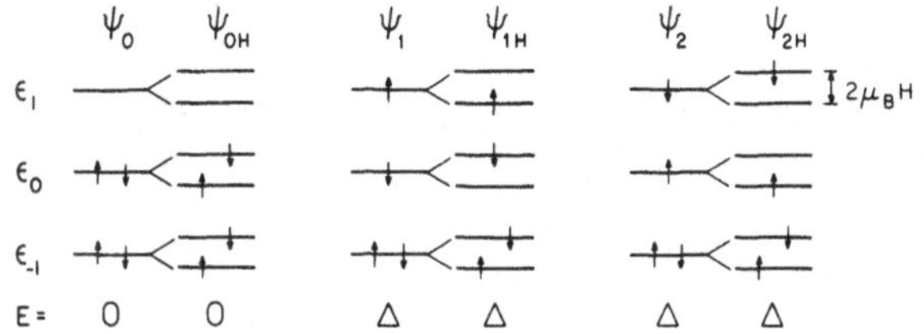

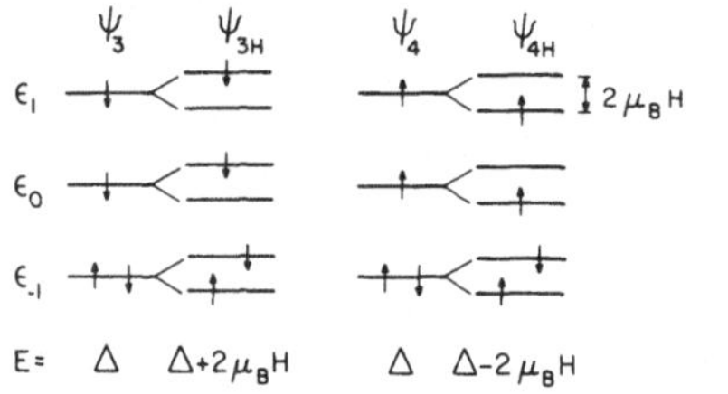

Fig. 2.9. Ground state and excited states of a microparticle with an even number of electrons in zero field (Ψ_i) and in a finite magnetic field (Ψ_{iH}). The corresponding energies E of the states measured with respect to the ground state are given

at T = 0, the electrons fill these levels up to ε_0. This leads to the ground state Ψ_0 depicted in Fig.2.9. At very low temperatures ($k_BT \ll \varepsilon_1-\varepsilon_0 = \Delta$) only the lowest lying excited states contribute significantly to the thermodynamic properties. For H = 0, there are four such states $\Psi_1,\ldots,\Psi_4$ which correspond to the promotion of an electron from the ε_0 to the ε_1 level. In Fig.2.9 we also show the excited states $\Psi_{1H},\ldots,\Psi_{4H}$ resulting from the application of an external field H. These will be used to calculate the spin paramagnetism of an even m.m.

In a zero magnetic field, the excited states $\Psi_1,\ldots,\Psi_4$ are of equal energy Δ (measured from the ground state), so that the partition function is

$$Z_{\mathrm{even}} = 1 + 4\exp(-\Delta/k_BT + \ldots) \quad . \tag{2.24}$$

The corresponding free energy reads

$$F_{\mathrm{even}} = -k_BT \ln Z_{\mathrm{even}} = -k_BT \ln[1 + 4\exp(-\Delta/k_BT + \ldots)] \simeq -4k_BT\exp(-\Delta/k_BT) \;, \tag{2.25}$$

where we use the assumption $\Delta \gg k_BT$.

In the presence of an external magnetic field, the states Ψ_{3H} and Ψ_{4H} have energies $\Delta + 2\mu_BH$ and $\Delta - 2\mu_BH$, respectively, where μ_B is the Bohr magneton. The states Ψ_{1H} and Ψ_{2H} have the original energy Δ. Consequently, the partition function is given by

$$Z_{H,\mathrm{even}} = 1 + 2\exp(-\Delta/k_BT) + \exp[-(\Delta+2\mu_BH)/k_BT] + \exp[-(\Delta-2\mu_BH)/k_BT] + \ldots \;. \tag{2.26}$$

This implies, for $k_BT \ll \Delta$, a free energy of the form

$$F_{H,\mathrm{even}} = -k_BT\{2\exp(-\Delta/k_BT) + \exp[-(\Delta+2\mu_BH)/k_BT] + \exp[-(\Delta-2\mu_BH/k_BT] + \ldots\} \;. \tag{2.27}$$

2.4.3 Free Energy of an "Odd" Particle

In a particle with an odd number of electrons there are two ground states and four excited states as shown in Fig.2.10 both for H = 0 and H > 0. For H = 0, the partition function becomes

$$Z_{odd} = 2 + 2\exp(-\Delta/k_BT) + 2\exp(-\Delta'/k_BT) + \dots \quad , \tag{2.28}$$

where $\Delta = \varepsilon_1 - \varepsilon_0$ and $\Delta' = \varepsilon_0 - \varepsilon_{-1}$. Assuming $k_BT \ll \Delta$, the resulting free energy is

$$\begin{aligned} F_{odd} &= -k_BT \ln\{2[1 + \exp(-\Delta/k_BT) + \exp(-\Delta'/k_BT) + \dots]\} \\ &= -k_BT \ln 2 - k_BT \ln\ [\exp(-\Delta/k_BT) + \exp(-\Delta'/k_BT) + \dots] \quad . \end{aligned} \tag{2.29}$$

In an external magnetic field the energies of the excited states are modified, as shown in Fig.2.10, leading to the partition function

$$Z_{H,odd} = [\exp(\mu_BH/k_BT) + \exp(-\mu_BH/k_BT)]\ [1 + \exp(-\Delta/k_BT) + \exp(-\Delta'/k_BT)] \quad . \tag{2.30}$$

This expression yields, for $k_BT \ll \Delta$, the free energy

$$\begin{aligned} F_{H,odd} &= -k_BT \ln\ [\exp(\mu_BH/k_BT) + \exp(-\mu_BH/k_BT)] \\ &\quad -k_BT\ [\exp(-\Delta/k_BT) + \exp(-\Delta'/k_BT)] \quad . \end{aligned} \tag{2.31}$$

We note that for the calculation of the magnetic moment $M = -\partial F/\partial H$, only the first terms of (2.31) contributes. This corresponds just to the magnetic moment of one extra unpaired electron independent of the level spacings Δ and Δ'.

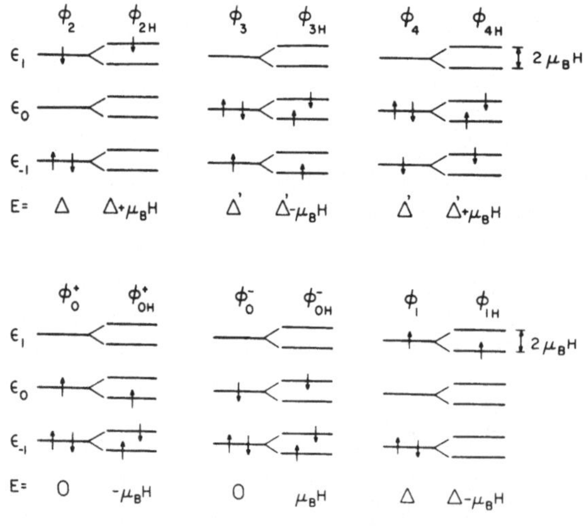

Fig. 2.10. Ground states and excited states of a microparticle with an odd number of electrons in zero field (ϕ_iH)

2.4.4 Thermodynamic Properties of a Single Metallic Microparticle

For the interpretation of the experiments on m.m., the thermodynamic properties derivable from the free energy expressions (2.24,25,27,31) must be properly averaged. The averaging process involves two steps [2.110]. In the first step one considers an ensemble of particles of the same size (measured by particle volume) but with different irregular shapes, leading to random variations of Δ. Averaging over the distribution $P(\Delta)$ of this ensemble yields the thermodynamic properties as a function of the average level spacing δ. The second step involves averaging of the latter quantities over the particle size distribution. This averaging procedure is appropriate for integrated quantities, such as the electronic heat capacity and the magnetic susceptibility measured by magnetometers. In the interpretation of the nuclear magnetic resonances (NMR) studies of the magnetic susceptibility, however, it is essential to bear in mind that particles with different moments have different NMR frequency shifts (Sect.2.4.7). Consequently, the distribution of the particle moments is reflected in the resulting line shape. For this case it is customary to consider first the average susceptibility over an ensemble of particles of variable shapes but of the same size and assume that particles of such an ensemble have the same moment, resulting in a sharp resonance line. The distribution of particle sizes is then considered as a source of line broadening [2.18]. However, we would like to point out that the irregular shape fluctuations may contribute to the line broadening as well, since the particle moment is [see(2.36)] a sensitive function of the "fluctuating" quantity Δ.

The *electronic heat capacity* is obtained from the general relation

$$C = -T \frac{\partial^2 F}{\partial T^2} \quad . \tag{2.32}$$

Introducing (2.25) into (2.32), we obtain for the heat capacity of an *even* m.m. with the level separation Δ

$$C_{\mathrm{even}}(\Delta) \simeq 4k_B(\Delta/k_BT)^2 \exp(-\Delta/k_BT) \quad . \tag{2.33}$$

We note that this result is valid for $\Delta \gg k_BT$. It is interesting that (2.33) agrees with the low-temperature heat capacity of a two-level system describing the Schottky anomaly for magnetic systems [2.105]. This is not surprising in view of the fact that all the lowest excited states (Fig.2.2) are of the same energy Δ. For an *odd* particle we use in (2.32) the expression (2.29) where, for simplicity, we put $\Delta = \Delta'$. The assumption $\Delta \gg k_BT$ then yields

$$C_{\mathrm{odd}}(\Delta) \simeq 2k_B(\Delta/k_BT)^2 \exp(-\Delta/k_BT) \quad . \tag{2.34}$$

The *magnetic moment* per particle is calculated from the relation

$$M = -\partial F_H/\partial H \quad . \tag{2.35}$$

For an *even* particle (2.27) imply, for $\mu_B H \ll k_B T$,

$$M_{even}(\Delta) \approx (8\mu_B^2/k_BT)\exp(-\Delta/k_BT)\,H \quad , \tag{2.36}$$

where we retain only the linear terms in H. We notice, for $\Delta \gg k_BT$, a strong suppression of the induced paramagnetic moment in an even particle via the factor $[\exp(-\Delta/k_BT)]$. For an *odd* particle the situation is different. Substitution of (2.31) into (2.35) yields

$$M_{odd}(\Delta) \approx (\mu_B^2/k_BT)\,H \quad . \tag{2.37}$$

This is just the Curie-like paramagnetic moment of a single unpaired electron. In very small particles this leads to a relatively large effect as compared to what one would obtain from the usual Pauli paramagnetism. The NMR consequences of these results are discussed in Sect.2.4.7.

2.4.5. Ensemble Averaging

The individual m.m. properties described above, such as $C_{even}(\Delta)$, $C_{odd}(\Delta)$, and $M_{even}(\Delta)$, must now be averaged over an appropriate distribution $P(\Delta)$. Assuming that the sample is a mixture of an equal amount of odd and even particles (of the same size), the ensemble-averaged electronic heat capacity is given by the ensemble average of (2.33) as follows:

$$C = (\langle C_{even}\rangle + \langle C_{odd}\rangle)/2 = 3k_B \int (\Delta/k_BT)^2 \exp(-\Delta/k_BT)\,P(\Delta)d\Delta \quad . \tag{2.38}$$

For the magnetic susceptibility it is necessary to distinguish the contribution of the even particles from the odd ones. The ensemble-averaged susceptibility of the even particles χ_{even} is calculated with the use of (2.36) as follows:

$$\chi_{even} = \langle M_{even}\rangle/H \approx (8\mu_B^2/k_BT)\int \exp(-\Delta/k_BT)\,P(\Delta)d\Delta \quad . \tag{2.39}$$

The susceptibility due to the odd particles is obtained from (2.37), viz

$$\chi_{odd} = \mu_B^2/k_BT \quad . \tag{2.40}$$

In order to evaluate the averages (2.38,39) the distribution function $P(\Delta)$ must be known. In KUBO's original work [2.109] the one-electron energy levels were described by a purely random process with the Poisson distribution

$$P(\Delta) = e^{-\Delta/\delta}/\delta \quad , \tag{2.41}$$

where δ is the average level spacing. According to (2.41), there is a finite probability of finding the value $\Delta = 0$ in the ensemble. However, two crossing energy eigenvalues tend to repel each other if there is a matrix element of the perturbing potential due to the irregular surface. This will modify the form of the distribution for small values of Δ. This point has been emphasized by GORKOV and ELIASHBERG [2.111], who suggested the analogy to the random matrix theory previously developed by WIGNER to discuss the nuclear spectra [2.112]. A very clear introduction to this subject is given in the review paper by KUBO [2.17]. We start with the simplest case of the *orthogonal* ensemble. The energy levels of such an ensemble follow from the eigenvalue problem

$$\det\begin{pmatrix} x-\varepsilon & y \\ y & -x-\varepsilon \end{pmatrix} = 0 \quad , \tag{2.42}$$

where $\pm x$ and y are the diagonal and the off-diagonal matrix elements of the random potential acting on the electrons of the m.m. of an irregular shape. Equation (2.42) implies the eigenvalues ε

$$\varepsilon = \pm\sqrt{x^2 + y^2} \quad . \tag{2.43}$$

Hence the spacing $\Delta = 2\varepsilon$ is a function of two independent random numbers x and y. To determine $P(\Delta)$, we must know $Pr(\Delta,\Delta+d\Delta)$, which is proportional to the area of the ring between the radii Δ and $\Delta+d\Delta$ in the x-y plane. Since this area is proportional to $\Delta \cdot d\Delta$, we have with the use of (2.21)

$$P(\Delta) = \text{const.}\ \Delta \quad . \tag{2.44}$$

We see that as a result of the level repulsion, caused by the random off-diagonal elements y, the distribution $P(\Delta) \to 0$ as $\Delta \to 0$. Other more complicated ensembles can be obtained by extending the above argument to include the effects of the spin-orbit interaction and the external magnetic field [2.17]. If one writes for small values of Δ a general expression of the form

$$P(\Delta) = \text{const.}\ \Delta^n \quad , \tag{2.45}$$

then the following different situations can be classified by the values of the integer n.

a) The above case of $n = 1$ (orthogonal ensemble) is valid for weak spin-orbit interaction and $H = 0$.

b) When a strong magnetic field is present, $n = 2$ is obtained (unitary ensemble) since Δ is a function of three independent random variables.

c) If the spin-orbit interaction is strong and $H = 0$, the Hamiltonian matrix [see (2.42)] is a 4 x 4 matrix, corresponding to Kramer's doublets. Consequently, Δ becomes a function of five independent variables leading to $n = 4$ (symplectic ensemble).

The Poisson distribution (2.42), classified as the random ($n = 0$) ensemble, would be obtained if the irregular potential had only diagonal matrix elements among the perfect particle states. That this is probably not the case is suggested by recent *ab initio* calculations of the eigenvalues of free electrons confined by an irregular surface leading to spectra which are closely approximated by the $n = 1$ ensemble [2.113].

2.4.6 Electronic Heat Capacity

Inserting (2.38) we find that the electronic heat capacity of a sample composed of particles of average level spacing δ reads

$$C(\delta) = \text{const.}\ \delta^{-(n+1)}\ T^{n+1} \quad . \tag{2.46}$$

Subsequent averaging of (2.46) over the size distribution does not affect the predicted temperature dependence, it only modifies somewhat the proportionality constant. From the experimental point of view, however, the actual numerical values of this constant are of interest. For example, consider the orthogonal ensemble for which DENTON et al. [2.110] find

$$C^{n=1}/k_B = 24(k_BT/\delta)^2 \quad . \tag{2.47}$$

We now compare this result with the bulk electronic heat capacity for the same number of electrons N [2.105]

$$C^{bulk}/k_B = \pi^2N(k_BT/2E_F) = 6.57(k_BT/\delta) \quad , \tag{2.48}$$

where we have used (2.22). The decrease of the electronic heat capacity of the m.m. compared with the bulk is then expressed by the ratio

$$C^{n=1}/C^{bulk} = 3.6(k_BT/\delta) \quad . \tag{2.49}$$

At very low temperatures ($k_BT \ll \delta$) we therefore expect a sizable reduction of the electronic heat capacity of the m.m. samples. The decrease is even more pronounced for $n > 1$, as indicated by (2.46). These results may explain why in certain m.m. the vibrational specific heat is dominant (Sect.2.22). STEWART [2.114] measured the heat capacity of sputtered granular metal films of Pt-SiO_2 at temperatures between 1 and 12 K. At the lowest temperatures (1.25 K) the sample with an average Pt-particle size of δ = 26 Å showed a decrease in the heat capacity, which supports (2.49). Other films, however, showed an enhancement of the heat capacity that was attributed to the vibrational modes of the structural defects induced by the presence of the Pt grains in the matrix. This is a typical example of a complication caused by the presence of the matrix supporting the m.m.

There are other examples for such a spurious contribution of the matrix. In magnetic resonance experiments a sizable line broadening takes place due to the impurity paramagnetic moments in the matrix [2.18]. Also the dielectric losses in the oxide coating of the m.m. seem to mask the genuine far-infrared absorption of the metallic cores (Sect.2.5.4) [2.115].

2.4.7. Spin Susceptibility and NMR Shift

The low-temperature spin susceptibility of an ensemble of even particles of a given size is obtained from (2.39,45). It varies with T as T^n. For the particular case of the *orthogonal* ensemble, the calculations of DENTON et al. [2.110] lead to

$$\chi_{even}^{n=1} = 7.63\ \mu_B^2(k_BT/\delta^2) \quad . \tag{2.50}$$

This result should be compared with the Pauli susceptibility [2.105] contributed by N electrons in a particle of level spacing δ

$$\chi_p = \mu_B^2(3N/2E_F) = 2\mu_B^2/\delta \quad , \tag{2.51}$$

where we used (2.22) to relate N to δ. Equations (2.50,51) imply

$$\chi_{even}^{n=1}/\chi_p = 3.81\ (k_BT/\delta) \quad . \tag{2.52}$$

For the odd particles, [2.40,51] yield

$$\chi_{odd}/\chi_p = \delta/2k_BT \quad . \tag{2.53}$$

These predictions have important implications for the nuclear magnetic resonance experiments. At a fixed NMR frequency the resonance of a nuclear spin is observed at a slightly different magnetic field in a sample of m.m. than in a diamagnetic solid. This effect is a kind of microparticle "Knight shift" [2.177] and it is characterized by [2.18]

$$K = \Delta H/H \quad , \tag{2.54}$$

where ΔH is the hyperfine field induced at the nuclear site by the electron polarization $\langle S_z \rangle$. Expressing the latter in terms of the electron spin susceptibility χ_e, we have

$$K = a\ I_z\ \chi_e/N\mu_N\mu_B g \quad , \tag{2.55}$$

where a is the hyperfine interaction constant, μ_N and I_z are the nuclear moment and spin, g is the electronic g-factor. As pointed out in Sect.2.4.4 the fluctuations of χ_e in an ensemble of particles of equal size are usually neglected so that one uses the average (2.39) for χ_e to calculate the Knight shift. Using (2.52,55) we obtain

$$K_{even}^{n=1}/K_p = 3.81\ (k_BT/\delta), \tag{2.56}$$

where K_p is the bulk Knight shift. For the odd particles (2.53) implies

$$K_{odd}/K_p = \delta/2k_BT\ . \tag{2.57}$$

Under the assumption of (2.22), a m.m. containing about 10^3 electrons has an average level spacing $\delta/k_B = 1.3 \times 10^2$K for $E_F/k_B = 10^5$K. Taking T = 1K, (2.56,57) yield a relative shift 2.86×10^{-2} and 66, respectively. Consequently, we expect to see NMR lines, one corresponding to even particles near a zero shift, the other due to odd particles with a large shift. The particle size distribution produces fluctuations of δ that are responsible for the broadening of the NMR lines [2.18].

TAUPIN [2.116] observed an unshifted NMR line coming from even particles of Li in LiF matrix. This is the first confirmation of the prediction $K_{even} \propto T$ as $T \to 0$ given by (2.56). KOBAYASHI et al. [2.117] studied NMR in oxide-coated Al particles prepared by gas evaporation. A sharp unshifted NMR line and a long tail were seen corresponding to even and odd particles, respectively. For particles of Cu produced in the same way, KOBAYASHI et al. [2.118] found that the zero-temperature value of K_{even} does not vanish. Subsequently, YEE and KNIGHT [2.119] studied Cu particles covered with SiO with a narrow size distribution and confirmed a finite value of K_{even} (T=0) proportional to δ^{-1}. Moreover they verified the linear T dependence of the finite temperature part of $K_{even}(T)$, indicating that the n = 1 ensemble may be an appropriate choice.

2.4.8 Spin-Orbit Coupling and Electron-Spin Resonance

The finite value of $K_{even}(T=0)$ is an indication of an inexact spin pairing in the ground state of the m.m. An analogous situation was found long ago in superconductors [2.120]. FERRELL [2.121] and ANDERSON [2.122] proposed that spin reversal of conduction electrons via spin-orbit coupling was responsible for this. DENTON et al. [2.110] suggested that a similar mechanism may act in even m.m. Detailed calculations along these lines were performed by SHIBA [2.123].

In the following we present a "pedestrian" derivation of $\chi_{even}(T=0)$ using the simplified model of Fig.2.9. The spin unpairing of the ground state is a result of the admixture of the triplet states Ψ_3 and Ψ_4 into Ψ_0 by the spin-orbit coupling H_{so}. In the presence of a magnetic field, the perturbed ground state is

$$\Psi_0' \simeq \Psi_0 + \frac{\langle 0|H_{so}|1\rangle}{\Delta+2\mu_BH}\,\Psi_{3,H} + \frac{\langle 0|H_{so}|1\rangle}{\Delta-2\mu_BH}\,\Psi_{4,H}\ . \tag{2.58}$$

The resulting spin polarization at T = 0 is then

$$\langle\Psi_0'|S_z|\Psi_0'\rangle = |\langle 0|H_{so}|1\rangle^2[(\Delta-2\mu_BH)^{-1} - (\Delta+2\mu_BH)^{-1}] \simeq 4\mu_BH|\langle 0|H_{so}|1\rangle|^2/\Delta^3\ . \tag{2.59}$$

SHIBA [2.123] takes into account all excited states that result from the assumed equidistant energy spectrum. This leads to a modification of (2.59) by a numerical factor and replacement of $|\langle 0|H|1\rangle|^2$ by an average $|\langle m|H_{so}|n\rangle|^2_{av}$. In this way SHIBA [2.123] obtains

$$\chi_{even}(T=0)/\chi_p \simeq |\langle m|H_{so}|n\rangle|^2_{av}/\delta^2 \quad . \tag{2.60}$$

Following the analogy to the superconductor [2.122], (2.60) can be expressed via the spin-reversal relaxation time τ_s of the conduction electron [2.124]. For a m.m. this is given by a golden-rule expression [2.123]

$$\hbar/\tau_s \simeq |\langle Km|H_{so}|n\rangle|^2_{av}/\delta \quad , \tag{2.61}$$

so that (2.60) can be written as

$$\chi_{even}(T=0)/\chi_p \simeq \hbar/\tau_s\delta \quad . \tag{2.62}$$

The measurements of $K_{even}(T=0)$ in Cu particles indicate $\tau_s = 5\times10^{-13}$s for particles of radius $R = 20$ Å [2.18]. This implies that the electron makes about 200 surface encounters per single spin reversal.

We note that (2.61) describes a real life-time process only if the particle is sufficiently large so that the electronic levels form a quasi-continuum spectrum. For a very small particle, the spin reversal relaxation process should be inhibited by the difficulty to match the initial- and final-states energies [2.17]. This is because the spin flip is accompanied by a small change of Zeeman energy, while the change of the orbital energy is of the order of $\delta \gg \mu_B H$. This effect is expected to lead to a narrowing of the line of the conduction electron-spin resonance (CESR) due to the unpaired electrons in odd m.m. KAWABATA [2.125] calculated the CESR linewidth $\delta\omega_e$ to be

$$\delta\omega_e/\omega_z \simeq \hbar/\tau_s\delta, \tag{2.63}$$

where ω_z is the resonance frequency $g\mu H/\hbar$. Hence CESR should be easier to observe in m.m. than in bulk metals. This expectation has found support in the observation of CESR in m.m. of gold [2.126], silver [2.127], and sodium [2.128]. The experiments of GORDON [2.128] suggest that the CESR is affected by the exchange interaction of the conduction electrons with the paramagnetic impurities in the matrix near the surface of the m.m. Also hyperfine interactions with nuclear spins in the m.m. are shown to play a role in the CESR linewidth [2.18,128]. A substantial reduction of the interactions with the matrix has been achieved by BOREL and MILLET [2.129] for m.m. of lithium dispersed in a xenon matrix.

KUBO [2.17] pointed out that also the *Korringa relaxation* process for nuclear spins should be quenched by the level quantization effect. Measurements of the nuclear-spin lattice relaxation time T_1 on aluminum and copper particles prepared by gas evaporation [2.117,118] did not show such a suppression. A level broadening caused by interparticle electron tunneling was proposed to account for the absence of the quenching effect [2.130]. More recently KOBAYASHI [2.131] succeeded in observing the enhancement of T_1 on ultrafine particles of Cu (15-25 Å in diameter) prepared by alternate evaporations of Cu and SiO to prevent oxidation [2.119].

Another effect, which may exhibit a similar sort of suppression in m.m., is the *Kondo* effect, which is known to occur in the presence of an antiferromagnetic exchange interaction between the impurity and the conduction electron spins [2.105]. Since an energy conserving spin flip of the latter spins is essential in this effect, we may expect a reduction of the Kondo singularity as a result of the quantization of the energy levels. This is also supported by the arguments of NOZIÈRES [2.132], who considered the Kondo effect in a finite crystal from the point of view of renormalization group theory. According to NOZIÈRES, the antiferromagnetic correlations between the impurity and conduction electron spin are confined within a coherence length $\xi \simeq \hbar v_F/T_k$ about the impurity, where T_k is the Kondo temperature. If the radius of the particle is much smaller than ξ, these correlations cannot develop and the Kondo effect should be quenched. This consideration could have some implication for experiments on magnetic susceptibility and resonance in samples involving impurity spins of the matrix adjacent to the m.m. Another possibility is the Kondo effect of the unpaired electrons in odd m.m. In general, one would expect a new effect to exist in m.m. due to the "unlocking of spins" which in the bulk are, for $T < T_k$, compensated by the Kondo correlations. Possible consequences may include enhanced nuclear spin relaxation, which is perhaps responsible for the anomalous shortening of T_1 observed in Al and Cu particles by KOBAYASHI et al. [2.117,118].

2.5 Electromagnetic Properties of Metallic Microparticles

In recent years there has been a renewed interest in various electromagnetic properties of metallic microparticles and their aggregates. This activity originated partly from the unusual predictions of the theoretical work of GORKOV and ELIASHBERG [2.111]. Section 2.5.1 is devoted to the predicted anomalous enhancement of the static electric polarizability of m.m. The resolution of the discrepancy between this prediction and experiments is reviewed in Sect.2.5.2. Section 2.5.3 introduces the classical problem of plasma resonance in spherical m.m. imbedded in a dielectric. In Sect.2.5.4 we discuss the experimental results for far-infrared absorption in aggregates of oxidized m.m. and their theoretical interpretations.

2.5.1 Electric Polarizability

We first discuss the prediction of an anomalously large electric polarizability caused by the level quantization [2.111]. To test this prediction MEIER and WYDER [2.133] measured the polarizability of small gold particles grown in a photosensitive glass. No anomalous polarizability was found down to temperatures of 0.3 K. A similar negative result was found by DUPREE and SMITHARD [2.134] for small silver particles. The discrepancy was subsequently resolved by STRÄSSLER and RICE [2.135], who pointed out an error in [2.111] caused by the neglect of the depolarization effects. In spherical particles the latter lead to the nonexistence of the predicted anomaly of electric polarizability [2.111]. The depolarization effect, however, should not arise in one-dimensional metals. This was pointed out by RICE and BERNASCONI [2.136], who predicted large enhancements of the dielectric constant of a quasi-one-dimensional metal formed by the mixed valency planar complex compounds of Pt. This idea has technological implications in the design of artificial dielectrics. The study of electric polarizability of m.m. is also of theoretical importance for the understanding of the Van der Waals interactions between particles [2.105].

2.5.2 Gorkov-Eliashberg Anomaly

The anomalous enhancement of the electric polarizability predicted in [2.111] is easy to derive. As in Sect.2.4.2 we regard the m.m. as a large molecule and calculate the electric dipole moment induced by the interaction zeE, where z is measured from the center of the m.m. and E denotes the electric field. This interaction mixes the ground state Ψ_0 of an even particle (Fig.2.9) with the excited states Ψ_1 and Ψ_2. Hence the perturbed ground state becomes

$$\Psi_0' \simeq \Psi_0 + \frac{e\langle 0|z|1\rangle E}{\Delta}(\Psi_1+\Psi_2) \quad . \tag{2.64}$$

Then the induced dipole moment is, to order E,

$$p = \langle\Psi_0'|ze|\Psi_0'\rangle \simeq \frac{4e^2}{\Delta}|\langle 0|z|1\rangle|^2 E \quad . \tag{2.65}$$

A more exact calculation of p involves all higher excited states [2.111], but the simple result (2.64) is adequate for the following discussion. Recalling the definition of the electric polarizability α_0

$$p = \alpha_0 E \tag{2.66}$$

and using the fact that the matrix element $\langle 0|z|1\rangle \simeq R$ where R is the radius of the m.m., we have from (2.65)

$$\alpha_0 \simeq \frac{e^2R^2}{\Delta} \quad . \tag{2.67}$$

It is well known from classical electrostatics that a metallic sphere of radius R has a polarizability

$$\alpha_m = R^3 \quad . \tag{2.68}$$

Equations (2.67,68) imply an enhancement of polarizability:

$$\frac{\alpha_0}{\alpha_m} \simeq \frac{e^2}{R\Delta} \simeq (k_F R)^2 \quad , \tag{2.69}$$

where we used (2.22) for Δ and assumed $me^2 \simeq \hbar k_F$, k_F being the Fermi wave vector. For $k_F R \gg 1$, (2.69) predicts large enhancements. Unfortunately, this one-electron calculation of α_0 also breaks down in this regime due to the interactions between the electrons occupying various "molecular" levels of the m.m. The main effect of these interactions is to screen the external field. In the classical approach to screening, one replaces (2.66) by the following:

$$p = \alpha_0 E_{loc} \quad ; \tag{2.70}$$

E_{loc} is the local field including the depolarization field [2.105]

$$E_{loc} = E - \frac{4\pi}{3} P \quad , \tag{2.71}$$

where P is the induced polarization (per unit volume). Equations (2.70,71) imply for the actual polarizability

$$\alpha = p/E = \alpha_0 / \left(1 + \frac{4\pi\delta_0}{3\Omega}\right) \quad , \tag{2.72}$$

where Ω is the volume of the particle. For macroscopic metallic spheres (2.69) implies $\alpha_0 \to \infty$, a divergence which can be attributed to the vanishing of the excitation energy Δ in a bulk metal. In this limit, (2.72) changes to $3\Omega/4\pi = R^3$, which agrees with the classical expression (2.68). This is the maximum polarizability obtainable for a spherical particle. For very small $R \sim k_F^{-1}$, the m.m. reduces to an "atom". In this limit (2.69) is expected to hold, but there is no enhancement. Thus (2.72) provides a simple explanation of the failure to observe the anomalous electric polarizability in m.m. [2.133,134].

2.5.3 Plasma Resonance Absorption

The plasma oscillation in a m.m. is a collective oscillation of the conduction electrons acted upon by a restoring force resulting from the induced charge on the surface. For a m.m. in vacuum the plasma resonance frequency can be found by first recalling the depolarization effect leading to (2.72). We note that for a spherical particle made of a metal of dielectric constant ε immersed in an infinite medium of dielectric constant ε_m, (2.72) is generalized to the following [2.137]

$$p/E = a^3 \left(\frac{\varepsilon-\varepsilon_m}{\varepsilon+2\varepsilon_m}\right) \quad . \tag{2.73}$$

In fact, (2.72) follows from (2.73) by letting $\varepsilon_m = 1$ and $\varepsilon = 1 + 4\pi\alpha_0/\Omega$. For a harmonically varying external field E of frequency ω the right-hand side of (2.73) is a function of ω. For the dielectric "constant" of the metal we use the "bulk" expression [2.105]

$$\varepsilon(\omega) = 1 - \frac{4\pi ne^2}{m\omega^2} = 1 - \frac{\omega_p^2}{\omega^2} \quad , \tag{2.74}$$

where n is the density of the conduction electrons and ω_p is the bulk plasma frequency. The spherical surface plasma resonance (s.p.r.) takes place when [2.138]

$$\varepsilon(\omega) + 2\varepsilon_m = 0 \quad , \tag{2.75}$$

for then (2.73) implies that the dynamic polarization $p/E \to \infty$. Using (2.74) in (2.75), we have for the s.p.r. frequency

$$\omega_0 = \omega_p/\sqrt{1+2\varepsilon_m} \quad . \tag{2.76}$$

The s.p.r. can be excited by light if the particle size is substantially smaller than the wavelength. For small gold particles in a SiO_2 matrix, (2.76) implies that the wavelength of the light giving rise to s.p.r. is about 5000 Å, much larger than the radius of the particles, which range between 10 and 100 Å [2.40-42]. DOREMUS [2.139,140] and HAMPE [2.141] measured the optical absorption of small particles of gold and found that the width of the s.p.r. is inversely proportional to the radius of the particle. This size effect can be explained by the limitation of the mean free path of the conduction electron, giving rise to an imaginary part of $\varepsilon(\omega)$. This amounts to replacing (2.74) by the Drude expression [2.138]

$$\varepsilon = 1 - \frac{\omega_p^2}{\omega^2+i\omega/\tau} \quad , \tag{2.77}$$

where $\tau \sim R/v_F$ is the electron lifetime. Extensive investigations of various size effects in the optical absorption by small gold particles were reported by KREIBIG [2.142]. On decreasing R, he also observed a diminishing of the temperature dependence of the spectra. This observation points towards the possibility of quenching the electron-phonon interaction in m.m. due to level quantization [2.143]. Moreover, the threshold energy of the interband transitions is found to decrease on decreasing R, which can be explained by changes of the lattice parameters [2.142].

2.5.4 Far-Infrared Absorption

For ultrafine metallic particles ($R \lesssim 200$ Å) the average level separation δ is comparable to the energy of the far-infrared photon. Under such circumstances the theory of GORKOV and ELIASHBERG [2.111] predicts a structure in the electromagnetic absorption due to the transitions between the discrete levels. In an effort to verify the predictions of [2.111], GRANQVIST et al. [2.144] studied the far-infrared absorption in powders of small oxidized Al particles. The structure was not observed, an observation that could be explained by the particle size distribution. A surprisingly large discrepancy was found between the strength of the measured absorptivity and the theory. For average particle radius less than 80 Å, GRANQVIST et al. [2.144] found an absorptivity that was well approximated by

$$\alpha(\tilde{\nu}) \simeq C_{exp}\,\tilde{\nu}^2 \quad , \tag{2.78}$$

where the inverse wavelength $\tilde{\nu}$ was in the region $3 < \tilde{\nu} < 150\ \mathrm{cm}^{-1}$. The proportionality constant C_{exp} was three orders of magnitude larger than that predicted either by GORKOV-ELIASHBERG [2.111] or Drude theory, see (2.77). We note that depolarization effects were taken into account in deriving the theoretical value of C_{theory} by using the MAXWELL-GARNETT theory of the dielectric constant of a collection of isolated spherical particles [2.145].

Another unexpected result was the absence of any observable changes due to superconducting ordering, such as a gap in the absorption, in Sn and Pb samples cooled below their transition temperatures [2.146]. We note that the particles studied in [2.146] were not small enough to cause a drastic smearing of the superconducting transition. This fact prompted investigations of alternative mechanism [2.115,147] not involving absorption by the metallic electrons. The mechanism of [2.115] is based on the observation that the particles tend to aggregate into tangled chains and clusters. This is expected to be caused by the strong van der Waals interaction between the particles. A strong absorption is predicted to occur in the amorphous oxide coating of the metallic particles because of the disorder-induced dielectric losses. Since the presence of metallic cores enhances the electric field in the oxide, the absorption also increases with the average radius R, as observed experimentally in [2.144]. A different explanation of the far-infrared anomaly [2.144] was proposed by GLICK and YORKE [2.147]. They considered direct excitation of phonon modes by the unscreened electric field acting on the ions at the surface of the m.m. This theory is able to explain the order of magnitude of the absorption, but it displays a decrease of absorption with R (as 1/R). The $\tilde{\nu}^2$ dependence of $\alpha(\tilde{\nu})$ in this theory results from assuming that bulk modes with density proportional to ω^2 are being excited. We note, however, that the vibrational modes of a microcrystal have a density, typically described by (2.6), involving a surface term proportional to ω. We expect this to modify the resulting $\tilde{\nu}$ dependence of the absorption calculated in [2.147].

2.6 Superconducting Properties

Superconducting ordering has been, so far, the most thoroughly studied cooperative effort in m.m. More precisely, one should talk about "attempted ordering" because of the fluctuations that dominate due to the small size of the particles. Section 2.6.1 contains an estimate of the mean-square thermodynamic order-parameter fluctuation as a function of the size of m.m. Sections 2.6.2-5 are devoted to various experimental manifestations of these fluctuations. The effects of particle size on the transition temperature are discussed in Sect.2.6.6.

2.6.1 Fluctuations of the Order Parameter

We note that the degree of superconducting order is quantitatively described in terms of a complex order parameter $\Psi(r,t)$, which represents the macroscopic wave function of the condensate of Cooper pairs [2.148]. For bulk superconductors, the temperature dependence of the average $\langle\Psi\rangle$ is well described by the BCS mean-field theory [2.148] in which $\langle\Psi\rangle$ vanishes above the transition temperature T_c (Fig.2.11). Actually, thermal fluctuations of $\Psi(r,t)$ take place, modifying this simple mean field picture. Above T_c, this corresponds to the temporary formation of regions of "condensed pairs" with an average size of the coherence length ξ. Fluctuations below T_c, on the other hand, are visualized as normal regions of size ξ on the background of the superconducting condensate. It is the very large value of ξ that distinguishes superconductivity from other types of phase transitions. For example, in pure aluminum we have $\xi(T = 0) = 1.6 \times 10^4$ Å. Because of their large size the fluctuating domains are associated with large changes of free energy that make the thermodynamic fluctuations very improbable events in bulk samples. This was first demonstrated quantitatively by GINZBURG [2.149] and THOULESS [2.150], who predicted that the specific heat of bulk superconductors is affected by fluctuations only in an extremely narrow temperature region of 10^{-15}K around the transition.

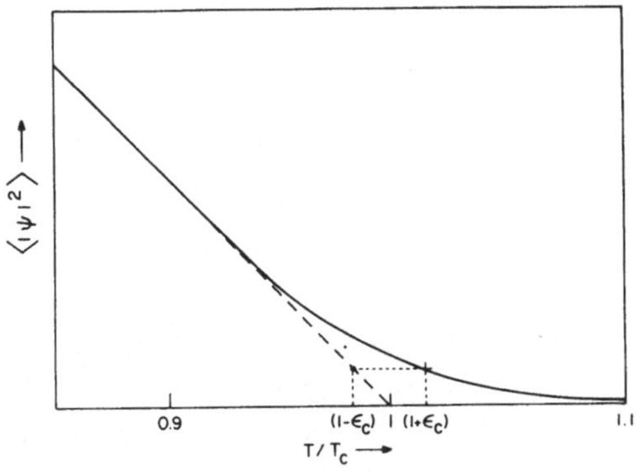

Fig. 2.11. Temperature dependence of the mean-square order parameter, given by (2.81), of an aluminum microparticle of radius R = 500 Å. The dotted line is the mean-field result for $|\psi|^2$ given by (2.82)

The above reasoning also suggests that the fluctuations can be made more probable by employing samples in which one or more dimensions are drastically reduced below the coherence length. About a decade ago this idea led to a flurry of interest in superconducting two-dimensional films and one-dimensional wires. Small superconducting particles, in which all dimensions are well below ξ, are called *zero-dimensional superconductors*. They provide an excellent tool for the study of thermodynamic order-parameter fluctuations since the condition $R \ll \xi$ is easily reached for standard values of the radius ($R \approx 10^2$ Å). Since the "fluctuating" order parameter is spatially constant in such particles, they are also convenient for theoretical investigations [2.151-155]. The thermodynamic properties, such as $\langle|\Psi|^2\rangle$ and the "fluctuation" specific heat, were first calculated by SCHMIDT [2.151] starting from the Ginzburg-Landau free-energy functional [2.149]

$$\mathcal{F}_{GL} = \Omega(\alpha|\Psi|^2 + \beta/2|\Psi|^4) \quad , \tag{2.79}$$

where Ω is the volume of the m.m., $\alpha \approx N_0(T-T_c)/T_c$, and $\beta \approx 0.1\ N_0/(k_BT_c)^2$, N_0 being the electron density of states at the Fermi level of the bulk metal. Using the expression

$$P(\Psi) = \exp(-\mathcal{F}_{GL}/k_BT)/\int \exp-\mathcal{F}_{GL}/k_BT)d^2\Psi \tag{2.80}$$

for the probability of a fluctuation, SCHMIDT [2.151] calculates the mean-square average of the order parameter:

$$\langle|\Psi|^2\rangle = \int \Psi^2 P(\Psi) d^2\Psi \quad . \tag{2.81}$$

The temperature dependence of $\langle|\Psi|^2\rangle$ near the bulk transition temperature T_c is shown in Fig.2.11. In Fig.2.11 we also show the mean-field value $|\Psi_0|^2$ that is obtained by minimizing $\mathcal{F}_{GL}$ with the result

$$\left.\begin{aligned} |\Psi_0|^2 &= -\alpha/\beta \quad &&\text{for } T \le T_c \\ &= 0 &&\text{for } T > T_c \end{aligned}\right\} \quad . \tag{2.82}$$

In contrast to the behavior of $|\Psi_0|^2$, which shows a discontinuous temperature derivative at T_c, the quantity $\langle|\Psi|^2\rangle$ exhibits a smooth temperature variation. This is consistent with the general result that systems with finite degrees of freedom do not exhibit phase transitions [2.156]. The degree of departure from the BCS behavior is characterized by the critical width ε_c. As indicated in Fig.2.11, ε_c is defined so that $\langle|\Psi|^2\rangle$ at $T/T_c = 1 + \varepsilon_c$ equals $|\Psi_0|^2$ at $T/T_c = 1 - \varepsilon_c$. An approximate expression for ε_c can be established by calculating $\langle|\Psi|^2\rangle$ from (2.81) where a "Gaussian approximation" is made for $P(\Psi)$ by taking $\mathcal{F}_{GL} \approx \Omega\alpha|\Psi|^2$. This leads to

$$<|\Psi|^2>_{gauss} = \frac{k_B T}{\Omega\alpha} \quad . \tag{2.83}$$

On comparing (2.83,82), we are led to the following expression

$$\varepsilon_c = \left(\frac{0.1}{\Omega N_0 k_B T}\right)^{1/2} \quad . \tag{2.84}$$

Recognizing that $(\Omega N_0)^{-1} = \delta$, we obtain form (2.84) the commonly used expression

$$\varepsilon_c \sim (\delta/k_B T_c)^{1/2} \quad . \tag{2.85}$$

Noting that rather narrow temperature regions ($\Delta T/T_c \sim 10^{-3}$) near T_c are experimentally accessible, we see from (2.85) that fluctuation effects should be detectable for average level spacings which are much smaller than $k_B T_c$. This is in contrast to the electron energy-level quantization effects which are usually observed only if $\delta \gg k_B T_c$ [2.131]. As the above result for $<|\Psi|^2>$ indicates, the basic effect of the fluctuations is a rounding of the phase transition over the region of temperatures given by (2.85). This "rounded" phase transition in zero-dimensional superconductors has been investigated by various methods, which are reviewed below.

2.6.2 Magnetic Susceptibility

For a bulk superconductor there is a complete expulsion of the magnetic flux leading to perfect diamagnetism characterized by a susceptibility $\chi = -1/4\pi$. Actually, the field penetrates a finite distance λ, called the London penetration depth, which is typically several hundred Å. For a m.m. this flux expulsion is only partial, and in the limit of $R \gg \lambda$ it is proportional to R^2/λ^2 [2.148]. Since λ^2 is proportional to the density of superconducting electrons $\sim|\Psi|^2$, the contribution of the m.m. to diamagnetism goes as $R^2|\Psi|^2$. This allows a detailed investigation of $|\Psi|^2$ as a function of temperature. BUHRMAN and HALPERIN [2.157] have studied such a diamagnetic transition for small Al particles with an average radius from 4000 Å to less than 125 Å. Their data show agreement with the Ginzburg-Landau calculations, including a suitable average over the size distribution.

2.6.3 Specific Heat

The fluctuation heat capacity can be calculated from the thermodynamic free energy $F = -k_B T \ln(Z)$ using the general expression (2.32). The partition function is in this case determined by the Ginzburg-Landau theory as

$$Z = \int \exp[-\mathcal{F}_{GL}(\Psi)/k_B T] d^2\Psi \quad . \tag{2.86}$$

The results of such a calculation show again a broadening of the mean-field jump over an interval of temperatures given by the critical width ε_c [2.151,158]. We note that the phenomenological theory based on the free-energy functional (2.79) allows only the superconducting fluctuations to be included in (2.86), the quasiparticle contribution being deleted. A more complete description of the thermodynamic properties of small superconductors was developed by MÜHLSCHLEGEL et al. [2.159]. These authors treat the BCS pairing interaction, in a discrete level system, with the use of the functional-averaging technique. This method incorporates into Z the quasiparticle contribution, which becomes increasingly important as T departs away from T_c [2.159]. WORTHINGTON et al. [2.160] have measured the heat capacity of granular aluminum films prepared by evaporation in the presence of oxygen. By increasing the oxygen pressure the oxide barrier becomes thicker and the grains become increasingly decoupled. Their results show a sharp BCS jump of the heat capacity for the film with the lowest resistivity ($\rho_N \simeq 6 \times 10^{-1}\,\Omega\,cm$). In the specimens with higher resistivity this jump exhibits a broadening, the degree of which increases with the resistivity. The measured specific heat curves resemble closely the theoretical results of [2.159], suggesting that the film can be regarded as a collection of independent Al grains with an effective radius increased by the electron transfer between the grains. We note that, as a result of this transfer, Josephson pair tunneling takes place which increases the size of the superconducting fluctuation regions beyond that of an isolated s.m.p. indicating a transition from zero- to three-dimensional superconductor [2.158].

2.6.4 Ultrasonic Attenuation

The attenuation of sound waves in normal metals is caused mainly by the conduction electrons. In a bulk superconductor this attenuation decreases rapidly as the sample is cooled below T_c. This is caused, first of all, by the decreasing number of quasiparticle excitations available to absorb the phonons. Also the coherence factors entering the matrix elements of the electron-phonon interaction contribute to this rapid temperature dependence [2.148]. In an attempt to see the effects of superconducting fluctuations on ultrasonic attenuation, ROBINSON et al. [2.161] have measured the attenuation of 2-GHz acoustic surface waves in a 300 Å aluminum film. In contrast to the above heat-capacity study [2.160], they have not seen any rounding of the BCS "break" in the temperature dependence of the attenuation at T_c. This can be attributed to the strong Josephson coupling which removes the fluctuation effects. Such an explanation is suggested by the low value of the resistivity of their film ($\rho_N \simeq 3 \times 10^{-1}\,\Omega\,cm$). In fact the heat-capacity measurements for films of similar resistivity show a sharp BCS jump [2.160]. From this point of view it would be worthwhile to extend the ultrasonic studies to films of higher resistivity. The potential of the ultrasonic attenuation method in small-particle physics is also emphasized by the work of TACHIKI et al. [2.162], who have shown that the Al_2O_3 matrix of the granular Al-film plays a rather secondary role in the absorption since the surface sound waves sample the Al-islands and the matrix in parallel.

2.6.5 Nuclear Spin Relaxation

The dominant mechanism for nuclear spin-lattice relaxation in metals is the contact hyperfine interaction between the conduction electrons and a nuclear magnetic moment [2.105]. The rate of the nuclear spin-lattice relaxation T_ℓ^{-1} is proportional in this case to the spectral density of the electron spin fluctuations. Expressing the latter, via the fluctuation-dissipation theorem, in terms of the local electron spin susceptibility $\chi(\underline{r},\omega_N)$ we have

$$T_\ell^{-1} \alpha (T/\omega_N) \; \mathrm{Im}\,\{\chi(\underline{r},\omega_N)\} \tag{2.87}$$

where ω_N is the nuclear magnetic resonance frequency. This result suggests that the nuclear relaxation time T_ℓ is a convenient local probe to study the effects of the superconducting fluctuations on the spin susceptibility of m.m. A theoretical estimate of such effects was first reported in [2.152,153], where the fluctuations of the order parameter Ψ were considered up to second order. A strong enhancement of the relaxation rate was predicted in the vicinity of the T_c [2.152,153]. The experimental results of KOBAYASHI et al. [2.117,163] on fine Al particles, however, did not confirm this prediction. In particular, the calculated rate appeared to be a gross overestimate of the fluctuation effects. This disagreement is linked to the divergence of the "Maki diagram", which represents a part of the calculated spin susceptibility [2.152,153]. We note that a similar difficulty was faced in the interpretation of the fluctuation effects upon the resistivity in one- and two-dimensional superconductors. KELLER and KORENMAN [2.164] and independently PATTON [2.165] have shown that this divergence can be removed by including the renormalization of the normal electrons due to their interaction with the superconducting fluctuations. A microscopic calculation of the nuclear relaxation rate in m.m., employing this renormalization procedure was performed in [2.154,155]. The results confirm the smooth variation of $1/T_\ell$ as a function of temperature observed previously by KOBAYASHI et al. [2.117]. The expression for the relaxation rate, valid in the vicinity of T_c, is of the form [2.115]

$$\frac{1}{T_\ell} \alpha\, T \int_{-\infty}^{\infty} \frac{df}{dx} \left[\left(\mathrm{Re}\left\{\frac{z}{\sqrt{z^2-\tilde{\Delta}^2}}\right\}\right)^2 + \left(\mathrm{Re}\left\{\frac{\tilde{\Delta}}{\sqrt{z^2-\tilde{\Delta}^2}}\right\}\right)^2\right] \tag{2.88}$$

where

$$z = x + i/2\tau_s \text{ and } 1/\tau_s = (8T/\pi)\tilde{\eta} \quad . \tag{2.89}$$

In (2.88), $f(x)$ is the Fermi function, τ_s is the order parameter relaxation time which is determined together with $\tilde{\Delta}^2$ from the coupled equations

$$\tilde{\eta} = \ln(T/T_c) + \frac{7\zeta(3)}{8}\frac{\tilde{\Delta}^2}{\pi T} \quad , \tag{2.90}$$

$$\tilde{\Delta}^2 = T/N_0\Omega\tilde{\eta} \quad . \tag{2.91}$$

Equations (2.90,91) correspond to the Hartree approximation for the superconducting fluctuation propagator [2.165]. From (2.88) we see that the main role of the finite size is to introduce a finite relaxation time τ_s which causes the rounding of the BCS quasiparticle density of states and of the $1/T_\ell$-versus- T-curve as well. As the size of the particle increases, we see from (2.89-91) that τ_s increases and expression (2.88) goes over to the bulk BCS limit [2.148]. This limit may also be reached by employing m.m. coupled by Josephson tunneling, as mentioned above. An improvement of the theory of [2.155] to extend its validity to temperatures well below T_c and to include the effects of external magnetic field was reported by SONE [2.166]. The measurements of T_ℓ on Al particles, by TAKAHASHI et al. [2.167], however, indicate that the theory of SONE [2.166] overestimates the effect of fluctuations. Subsequently, WARD et al. [2.168] proposed that Josephson coupling between particles could be responsible for a suppression of thermodynamic fluctuations. This is plausible because of the strong Van der Waals forces causing the m.m. to form tightly bound necklaces and clusters. The theory of WARD et al. [2.168] also explains the difference between the behavior of Sn and Al particles observed by KOBAYASHI et al. [2.169]. The Sn-particles are, as a rule, covered by a thinner layer of oxide than Al particles, producing a stronger Josephson coupling and a $1/T_\ell$-versus-T curve more closely resembling the BCS result [2.148].

2.6.6 Transition Temperature

As emphasized above, the thermodynamic fluctuations in an isolated small superconductor do not allow for a real phase transition at a defined temperature. However, a collection of m.m. coupled by Josephson tunneling exhibits a phase transition at a temperature which increases with the strength of tunneling up to a value which depends mainly on the electronic properties of the m.m. itself. It is the latter value which one usually denotes as the transition temperature of a small superconductor T_{cs}. We may define it loosely as a mean-field transition temperature of a "fictitious bulk" superconductor with the electronic properties of the small superconductor. There are two effects which contribute to the change of T_{cs} with respect to T_c of a "real" bulk superconductor. 1) The electronic level quantization effect decreases T_{cs} and eventually prevents ordering [2.170,171]. 2) The modification of the attractive interaction between electrons can either increase or decrease T_{cs}. It is convenient to discuss these effects starting from the self-consistency condition of the mean-field order parameter (gap equation of [2.171]),

$$1 = \frac{V}{\Omega} \sum_m \frac{\tanh(\varepsilon_m/2k_BT_{cs})}{\varepsilon_m} \tag{2.92}$$

where V is the attractive interaction constant. The summation goes over the one-electron levels ε_m (measured with respect to E_F) up to $\hbar\omega_D$, where ω_D is the Debye frequency. As the separation between the successive levels δ increases the number of terms contributing to the m-sum decreases. STRONGIN et al. [2.171] have shown that for

$$\delta \gtrsim 2\pi k_BT_c \tag{2.93}$$

the sum is too small to satisfy the condition (2.92), implying the absence of superconducting order. We note that T_c in (2.93) is the transition temperature of a bulk superconductor given by the BCS expression [2.148]

$$T_c = (1.14\hbar\omega_D/k)\exp(-1/N_0V) \quad . \tag{2.94}$$

The condition (2.93), for a "critical" value of δ, was predicted a long time ago by ANDERSON [2.170]. For an isolated lead particle it implies a critical radius of 22 Å, or about 750 electrons needed to give rise to a spontaneous superconducting order parameter. We note that GIAVER and ZELLER [2.172], using tunneling, have observed signs of superconducting ordering in particles of Sn down to 25 Å radius.

As seen in (2.94) the attractive interaction V plays an important role in T_c. In small metallic particles the parameter V is expected to be modified by the presence of a free surface or metal-insulator interface. In fact, thin films or grains having a high surface-to-volume ratio often show a higher transition temperature than bulk samples. The most widely accepted explanation for these observations is the softening of the surface phonons [2.7,173], leading to an enhancement of the phonon-mediated interaction between electrons (Fig.2.4). Coating the metallic particles with rare-gas solids leads to a decrease of T_{cs} which can be explained by a modification of the surface phonons [2.174]. An anomaly, in this respect, is the result of DEUTSCHER and PASTERNAK [2.175] showing a significant increase of T_c for hydrogen-coated Al film. This result points to a certain role of the chemical bonds (such as $Al\text{-}H_2$) in the superconducting pairing interaction [2.176].

Acknowledgements. It is a pleasure to thank Dr. K. Schürmann (Landis & Gyr) for a critical reading of this paper. The manuscript was typed by Mrs. Gabriella Baltes and Mrs. Glenna Paschal with patience and competence. H.P.B. gratefully acknowleges the hospitality of the Department of Physics at the University of California, Riverside.

References

2.1 H.B. Callen: *Thermodynamics* (Wiley, New York, London 1960)
2.2 W.H. Marlow: "Introduction - The Domains of Aerosol Physics", in: *Aerosol Microphysics, I: Particle Interactions*, ed. by W.H. Marlow, Topics in Current Physics, Vol. 16 (Springer, Berlin, Heidelberg, New York 1980) pp. 1-14
2.3 H.P. Baltes: J. Phys. (Paris) *38*, C-2, 151-156 (1977)
2.4 G.H. Comsa, D. Heitkamp, H.S. Räde: Solid State Commun. *24*, 547-550 (1977)
2.5 H.P. Baltes, E.R. Hilf: Solid State Commun. *12*, 369-373 (1973)
2.6 F. Meier, P. Wyder: Phys. Rev. Lett. *30*, 181-184 (1973)
2.7 V. Novotny, P.P.M. Meincke: Phys. Rev. B *8*, 4168-4199 (1973)
2.8 Ph. Buffat, J.-P. Borel: Phys. Rev. A *13*, 2287-2298 (1976)
2.9 C. Solliard, Ph. Buffat: J. Phys. (Paris) *38*, C2, 167-170 (1977)
2.10 M. Gillet: Surf. Sci. *67*, 139-157 (1977)
2.11 H.P. Baltes: Am. J. Phys. *42*, 505-507 (1974)
2.12 H.P. Baltes, B. Steinle, M. Pabst: Phys. Rev. A *13*, 1866-1873 (1976)
2.13 J.L. Lebowitz, J. Sykes: J. Stat. Phys. *6*, 157-171 (1972)
2.14 H.P. Baltes, E.R. Hilf: *Spectra of Finite Systems* (Bibliographisches Institut, Mannheim 1976)
2.15 International Meeting on the Small Particles and Inorganic Clusters, J. Phys. (Paris) *38*, Colloque C-2, Suppl. to No. 7 (1977)
2.16 J.A. Barker: J. Phys. (Paris) *38*, C2, 37-45 (1977)
2.17 R. Kubo: J. Phys. (Paris) *38*, C2, 69-75 (1977)
2.18 W.D. Knight: J. Phys. (Paris) *38*, C2, 109-113 (1977)
2.19 M. Hoare: Adv. Chem. Phys. (in press)
2.20 H. Müller-Krumbhaar: in *Monte Carlo Methods in Statistical Physics*, ed. by K. Binder, Topics in Current Physics, Vol. 7 (Springer, Berlin, Heidelberg, New York 1979) pp. 195-223
2.21 K. Binder, D. Stauffer: Adv. Phys. *25*, 343-396 (1976)
2.22 H.P. Baltes: Infrared Phys. *16*, 1-8 (1976)
2.23 H.P. Baltes: Appl. Phys. *72*, 221-244 (1977) Sect. 2
2.24 R. Kälin, F. Kneubühl: Infrared Phys. *16*, 491-508 (1976)
2.25 K.L. Kliewer, R. Fuchs: Adv. Chem. Phys. *27*, 355-541 (1974)
2.26 L. Genzel: Festkörperprobleme *14*, 183-203 (1974)
2.27 R. Ruppin, R. Englman: Rep. Progr. Phys. *33*, 149-196 (1970)
2.28 N.C. Wickramasinghe, K. Nandy: Rep. Progr. Phys. *35*, 157-234 (1972)
2.29 D.R. Huffman: Adv. Phys. *26*, 129-230 (1977)
2.30 F.H. Brownell: J. Math. Mech. *6*, 119-166 (1957)
2.31 R.K. Pathria: Nuovo Cimento, [Suppl. Ser. I, *4*, 276-288 (1966)]
2.32 R. Balian, C. Bloch: Ann. Phys. (New York) *60*, 401-447 (1970)
2.33 R. Balian, C. Bloch: Ann. Phys. (New York) *69*, 76-160 (1972)
2.34 R. Balian, C. Bloch: Ann. Phys. (New York) *84*, 559-563 (1974)
2.35 H.P. Baltes: Phys. Rev. A *6*, 2252-2257 (1972)
2.36 A.N. Chaba, R.K. Pathria: Phys. Rev. A *8*, 3264-3265 (1973)
2.37 H.P. Baltes, B. Steinle: J. Math. Phys. *18*, 1275-1276 (1977)
2.38 A.N. Chaba: Phys. Rev. A *20*, 1292-1294 (1979)
2.39 H.P. Baltes, F.K. Kneubühl: Helv. Phys. Acta *45*, 481-529 (1972)
2.40 B. Steinle, H.P. Baltes, M. Pabst: Phys. Rev. A *12*, 1519-1524 (1975)
2.41 B. Steinle, H.P. Baltes, M. Pabst: Infrared Phys. *16*, 25-27 (1976)
2.42. R. Balian, C. Bloch: Ann. Phys. (New York) *64*, 271-307 (1971)
2.43 R. Balian, C. Bloch: Ann. Phys. (New York) *104*, 300-335 (1977)
2.44 R. Stratton: Philos. Mag. *44*, 519-532 (1953)
2.45 M. Dupuis, R. Mazo, L. Onsager: J. Chem. Phys. *33*, 1452-1461 (1960)
2.46 R. Stratton: J. Chem. Phys. *37*, 2972-2974 (1962)
2.47 A.A. Maradudin, R.F. Wallis: Phys. Rev. *148*, 945-961 (1966)
2.48 M.G. Burt: J. Phys. C *6*, 855-867 (1973)
2.49 M. Planck: *Theorie der Wärmestrahlung*, 4th ed. (Barth, Leipzig 1921) p. 218
2.50 C. Schaefer: Z. Phys. *7*, 287-296 (1921)
2.51 K.H. Rieder: Surf. Sci. *26*, 637-648 (1971)
2.52 R. Lautenschläger: Solid State Commun. *16*, 1331-1334 (1974)

2.53 Th.F. Nonnenmacher: Phys. Lett. *51A*, 213-214 (1975)
2.54 V. Novotny, P.P.M. Meincke, J.H.P. Watson: Phys. Rev. Lett. *28*, 901-903 (1972)
2.55 Y. Kashiwase, I. Nishida, Y. Kainuma, K. Kimoto: J. Phys. (Paris) *38*, C2, 157-160 (1977)
2.56 K. Oshima: J. Phys. (Paris) *38*, C2, 163-165 (1977)
2.57 Ph.-A. Buffat: Solid State Commun. *23*, 547-550 (1977)
2.58 M.P.A. Viegers, J.M. Trooster: Phys. Rev. B *15*, 72-83 (1977)
2.59 D. Bijl: Philos. Mag. *43*, 1342-1344 (1952)
2.60 K. Case, S.C. Chiu: Phys. Rev. A *1*, 1170-1174 (1970)
2.61 H.P. Baltes, F.K. Kenubühl, R. Muri: in *Submillimeter Waves*, Vol. 20, ed. by J. Fox (Polytechnic Press, Brooklyn 1971) p. 667-691
2.62 H.P. Baltes, F.K. Kneubühl: Opt. Commun. *4*, 9-12 (1971)
2.63 H.P. Baltes: Appl. Phys. *1*, 39-43 (1973)
2.64 H.P. Baltes, E.R. Hilf, M. Pabst: Appl. Phys. *3*, 21-29 (1974)
2.65 W. Eckhardt: Opt. Commun. *14*, 95-98 (1975)
2.66 W. Eckhardt: Z. Phys. B *23*, 213-219 (1976)
2.67 W. Eckhardt: Z. Phys. B *26*, 291-297 (1977)
2.68 W. Eckhardt: Z. Naturforsch. *34a*, 292-299 (1979)
2.69 R.C. Powell: Phys. Rev. B *2*, 2090-2097 (1970)
2.70 R. Balian, B. Duplantier: Ann. Phys. (New York) *112*, 165-208 (1978)
2.71 G.S. Agarwal: Phys. Rev. Lett. *32*, 703-706 (1974)
2.72 H.P. Baltes, J. Geist, A. Walther: in *Inverse Source Problems in Optics*, ed. by H.P. Baltes, Topics in Current Physics, Vol. 9 (Springer, Berlin, Heidelberg, New York 1978) pp. 119-154
2.73 F.A. Meier: "Small Particles and Boundary Conditions", Ph.D. Thesis, Fakulteit der Wiskunde en Natuurwetenschappen, Katholieke Universiteit Nijmegen (1973)
2.74 G. Magens: "Spezifische Wärme und Magnetisierung kleiner Metallpartikel", Diplomarbeit, Fachbereich Physik, Freie Universität Berlin (1974)
2.75 E. Hilf: Z. Naturforschung *25a*, 1190-1202 (1970)
2.76 F.F. Abraham, R.N. Kortzeborn: J. Chem. Phys. *57*, 1985-1989 (1972)
2.77 G. Nenciu: J. Stat. Phys. *7*, 119-130 (1973)
2.78 R. McFee: Am. J. Phys. *41*, 230-234 (1973)
2.79 S. Greenspoon, R.K. Pathria: Phys. Rev. A *9*, 2103-2110 (1974)
2.80 A.N. Chaba, R.K. Pathria: Phys. Rev. A *18*, 1277-1281 (1978)
2.81 C. Domb: J. Phys. A *6*, 1296-1305 (1973)
2.82 K. Binder: Physica *62*, 508-526 (1972)
2.83 B. Stoeckly: Phys. Rev. B *14*, 1271-1275 (1976)
2.84 R. Nossal: J. Math. Phys. *6*, 193-200 (1965)
2.85 R. Nossal: J. Chem. Phys. *45*, 1097-1100 (1966)
2.86 I. Oppenheim, P. Mazur: Physica *30*, 1833-1845 (1964)
2.87 P. Mazur: Physica Norvegica *5*, 291-295 (1971)
2.88 K.H. Rieder: Phys. Rev. Lett. *20*, 209-211 (1968)
2.89 K.H. Rieder, W. Drexel: Phys. Rev. Lett. *34*, 148-151 (1975)
2.90 H. Fröhlich: *Theory of Dielectrics* (Oxford University Press, Oxford 1949) p. 149
2.91 R. Englman, R. Ruppin: J. Phys. C *1*, 614-643 (1968)
2.92 R. Fuchs, K.L. Kliewer: J. Opt. Soc. Am. *58*, 319-330 (1968)
2.93 R. Fuchs: Phys. Rev. B *11*, 1732-1740 (1975)
2.94 L. Genzel, T.P. Martin: Surface Sci. *34*, 33-39 (1973)
2.95 V.V. Bryksin, Yu.M. Gerbstein, D.N. Mirlin: Solid State Commun. *9*, 669-673 (1971)
2.96 L. Novák: Phys. Stat. Sol. (b) *56*, 307-312 (1973)
2.97 J. Nahum, R. Ruppin: Phys. Stat. Sol. (a) *16*, 459-462 (1973)
2.98 P. Clippe, R. Evrard, A.A. Lucas: Phys. Rev. B *14*, 1715-1721 (1976)
2.99 T.P. Martin: Phys. Rev. B *1*, 3480-3488 (1970); Phys. Rev. B *15*, 4071-4076 (1977)
2.100 T.R. Steyer, K.L. Day, D.R. Huffman: Appl. Opt. *13*, 1586-1590 (1974)
2.101 R. Kälin, H.P. Baltes, F.K. Kneubühl: Helv. Phys. Acta *43*, 487-488 (1970); Solid State Commun. *8*, 1495-1497 (1970)

2.102 T.S. Sun, A. Anderson: Spectrosc. Lett. *4*, 377-383 (1971)
2.103 E.S. Etz, G.J. Rosasco, J.J. Blaha: in *Environmental Pollutants*, ed. by Y. Taft, J.R. Coleman, B.E. Dahneke, I. Feldman (Plenum, New York 1978) pp. 413-456
2.104 M. Kerker, P.J. McNulty, M. Sculley, H. Chew, D.D. Cooke: J. Opt. Soc. Am. *68*, 1676-1689 (1978)
2.105 C. Kittel: *Introduction to Solid State Physics* (John Wiley & Sons, New York 1971)
2.106 L.D. Landau, E.M. Lifshitz: *Statistical Physics*, 2nd ed. (Pergamon, Oxford 1968)
2.107 H. Fröhlich: Physica *6*, 406-412 (1937)
2.108 D.A. Greenwood, R. Brout, J.A. Krumhansl: Bull. Am. Phys. Soc. *5*, 297 (1960)
2.109 R. Kubo: J. Phys. Soc. Jpn. *17*, 975-986 (1962)
2.110 R. Denton, B. Mühlschlegel, D.J. Scalapino: Phys. Rev. B *7*, 3589-3606 (1973)
2.111 L.P. Gorkov, G.M. Eliashberg: Zh. Eksp. Teor. Fiz. *48*, 1407-1418 (1965) [English transl.: Sov. Phys. JETP *21*, 940-947 (1965)]
2.112 E.P. Wigner: Proc. Cambridge Philos. Soc. *47*, 790-796 (1951)
2.113 J.F. Tavel, K.F. Ratcliff, N. Rosenzweig: Phys. Lett. *73A*, 353-355 (1979)
2.114 G.R. Stewart: Phys. Rev. B *15*, 1143-1150 (1977)
2.115 E. Šimànek: Phys. Rev. Lett. *38*, 1161-1163 (1977)
2.116 C. Taupin: J. Phys. Chem. Solids *28*, 41-47 (1967)
2.117 S. Kobayashi, T. Takahashi, W. Sasaki: J. Phys. Soc. Jpn. *31*, 1442-1443 (1971)
2.118 S. Kobayashi, T. Takahashi, W. Sasaki: J. Phys. Soc. Jpn. *32*, 1234-1236 (1971)
2.119 P. Yee, W.D. Knight: Phys. Rev. B *11*, 3261-3267 (1975)
2.120 G.M. Androes, W.D. Knight: Phys. Rev. Lett. *2*, 386-387 (1959)
2.121 R.A. Ferrell: Phys. Rev. Lett. *3*, 262-265 (1959)
2.122 P.W. Anderson: Phys. Rev. Lett. *3*, 325-326 (1954)
2.123 H. Shiba: J. Low Temp. Phys. *22*, 105-119 (1976)
2.124 R.J. Elliott: Phys. Rev. *96*, 266-279 (1954)
2.125 A. Kawabata: J. Phys. Soc. Jpn. *29*, 902-911 (1970)
2.126 R. Monot, A. Châtelain, J.P. Borel: Phys. Lett. *34A*, 57-58 (1971)
2.127 R. Monot, C. Narbel, J.-P. Borel: Nuovo Cimento *19B*, 253-260 (1974)
2.128 D.A. Gordon: Phys. Rev. *B13*, 3738-3747 (1976)
2.129 J.-P. Borel, J.-L. Millet: J. Phys. (Paris) *38*, C2, 115-119 (1977)
2.130 E. Šimánek: J. Phys. (Paris) *38*, C2, 79-82 (1977)
2.131 S. Kobayashi: J. Phys. (Paris) *38*, C2, 121-123 (1977)
2.132 P. Nozières: in *14th Low Temperature Physics Conf.*, ed. by M. Krusius, M. Vuorio (North Holland, Amsterdam 1975) pp. 339-374
2.133 F. Meier, P. Wyder: Phys. Lett. *39A*, 51-52 (1972)
2.134 R. Dupree, M.A. Smithard: J. Phys. C *5*, 408-414 (1972)
2.135 S. Strässler, M.J. Rice: Phys. Rev. B *6*, 2575-2577 (1972)
2.136 M.J. Rice, J. Bernasconi: Phys. Rev. Lett. *29*, 113-116 (1972)
2.137 J.D. Jackson: *Classical Electrodynamics*, 2nd ed. (John Wiley & Sons, New York 1975)
2.138 A. Kawabata, R. Kubo: J. Phys. Soc. Jpn. *21*, 1765-1772 (1966)
2.139 R.H. Doremus: J. Chem. Phys. *40*, 2389-2396 (1964)
2.140 R.H. Doremus: J. Chem. Phys. *42*, 414-417 (1965)
2.141 W. Hampe: Z. Phys. *152*, 476-494 (1958)
2.142 U. Kreibig: J. Phys. (Paris) *38*, C2, 97-103 (1977)
2.143 R. Kubo: Comments Solid State Phys. *1*, 61-64 (1968)
2.144 C.G. Granqvist, R.A. Buhrman, J. Wyns, A.J. Sievers: Phys. Rev. Lett. *37*, 625-629 (1976)
2.145 J.C. Maxwell-Garnett: Philos. Trans. Roy. Soc. London *203*, 385-420 (1904); *205*, 237-288 (1906)
2.146 D.B. Tanner, A.J. Sievers, R.A. Buhrman: Phys. Rev. B *11*, 1330-1341 (1975)
2.147 A.J. Glick, E.D. Yorke: Phys. Rev. B *18*, 2490-2493 (1978)
2.148 M. Tinkham: *Introduction to Superconductivity* (McGraw Hill, New York 1975)
2.149 V.L. Ginzburg: Fiz. Tverd. Tela 2031-2043 (1960) [English transl.: Sov. Phys. Solid State *2*, 1824-1834 (1960)]
2.150 D.J. Thouless: Ann. Phys. (New York) *10*, 553-588 (1960)

2.151 V.V. Schmidt: in *11th Int. Conf. on Low Temperature Physics, Proc.*, ed. by M.P. Malkow (Moscow Publishing House, Moscow 1967) pp. 205-209
2.152 K. Maki, J.P. Hurault, M.T. Beal-Monod: Phys. Lett. *31A*, 526-527 (1970)
2.153 J.P. Hurault, K. Maki, M.T. Beal-Monod: Phys. Rev. B *3*, 762-768 (1971)
2.154 E. Šimànek, D.E. MacLaughlin, D. Imbro: Phys. Lett. *42A*, 357-358 (1973)
2.155 E. Šimànek, D. Imbro, D.E. MacLaughlin: J. Low Temp. Phys. *11*, 787-802 (1973)
2.156 K. Huang: *Statistical Mechanics* (John Wiley & Sons, New York, London 1963)
2.157 R.A. Buhrman, W.P. Halperin: Phys. Rev. Lett. *30*, 692-695 (1973)
2.158 G. Deutscher, Y. Imry, L. Gunther: in *13th Int. Conf. on Low-Temperature Physics, Proc.*, ed. by K.D. Timmerhaus, W.J. O'Sullivan, E.F. Hammel (Plenum, New York 1974) pp. 692-695
2.159 B. Mühlschlegel, D.J. Scalapino, R. Denton: Phys. Rev. B *6*, 1767-1777 (1972)
2.160 T. Worthington, P. Lindenfeld, G. Deutscher: Phys. Rev. Lett. *41*, 316-319 (1978)
2.161 D.A. Robinson, K. Maki, M. Levy: Phys. Rev. Lett. *32*, 709-712 (1974)
2.162 M. Tachiki, H. Salvo, D.A. Robinson, M. Levy: Solid State Commun. *17*, 653-656 (1975)
2.163 S. Kobayashi: in *13th Low Temperature Phys. Conf. Proc.*, ed. by K.D. Timmerhaus, W.J. O'Sullivan, E.F. Hammel (Plenum, New York 1974) pp. 315-319
2.164 J. Keller, V. Korenman: Phys. Rev. Lett. *27*, 1270-1273 (1971)
2.165 B. Patton: Phys. Rev. Lett. *27*, 1273-1276 (1971)
2.166 J. Sone: J. Low Temp. Phys. *23*, 699-723 (1976)
2.167 T. Takahashi, S. Kobayashi, W. Sasaki: Solid State Commun. *17*, 681-684 (1975)
2.168 R.C. Ward, S. Cremer, E. Šimànek: Phys. Rev. B *17*, 3541-3546 (1978)
2.169 S. Kobayashi, T. Takahashi, W. Sasaki: J. Phys. Soc. Jpn. *36*, 714-719 (1974)
2.170 P.W. Anderson: J. Phys. Chem. Solids *11*, 26-30 (1959)
2.171 M. Strongin, R.S. Thompson, O.F. Kammerer, J.E. Crow: Phys. Rev. B *1*, 1078-1091 (1970)
2.172 I. Giaver, H.R. Zeller: Phys. Rev. Lett. *20*, 1504-1507 (1968)
2.173 S. Akselrod, M. Pasternak, S. Bukspan: J. Low Temp. Phys. *17*, 375-383 (1974)
2.174 D.G. Naugle, J.W. Barker, R.E. Allen: Phys. Rev. B *7*, 3028-3037 (1973)
2.175 G. Deutscher, M. Pasternak: Phys. Rev. B *10*, 4042-4043 (1974)
2.176 E. Šimànek: Solid State Commun. *32*, 731-733 (1979)
2.177 C.P. Slichter: *Principles of Magnetic Resonance*, Springer Series in Solid-State Sciences, Vol. 1 (Springer, Berlin, Heidelberg, New York 1980)

3. Electronic Structure Studies of Overlayers Using Cluster and Slab Models

I. P. Batra

With 14 Figures

The past decade has witnessed a tremendous amount of effort in theoretically understanding the electronic structure of surfaces. This effort has been in response to the major breakthroughs which were made in preparation and characterization of clean and absorbate-covered surfaces. Many spectroscopic techniques are now available to study a wide range of surface properties. Some of the examples of these properties include surface electronic structure, chemical bond formation, surface vibrations, and geometrical arrangement of surface atoms. Various current spectroscopic techniques are briefly summarized in Appendix 3.A.

In this chapter, we describe how clusters and slabs are used as models for calculating the electronic structure of adsorbate-covered surfaces. A particular example to be discussed is the oxygen(adsorbate)/aluminum(surface) system. The cluster and slab model calculation results are compared with the data obtained in photoemission spectroscopy, which is a technique for the investigation of electronic structure. In photoemission experiments a sample is irradiated with photons of appropriate energy, knocking out electrons which are then analyzed to obtain information about the energy levels from which these electrons originate. If the emitted electrons belong to overlayers, then the information obtained is relevant to the electronic structure of the adsorbate-substrate system.

Theoretical calculations of the electronic structure are complicated in general, but for surfaces, additional complexity arises due to the loss of translational invariance symmetry. There are many good articles in the literature on surface electronic calculations and only a representative list appears here [3.1-20]. To cope with the lack of the translational symmetry in the direction perpendicular to the surface, several theoretical models have been developed to treat the semiinfinite system. Some examples are the

a) scattering-theoretic technique [3.1,17],

b) transfer-matrix Green's function method [3.9,19-20],

c) density-functional jellium model [3.5], and

d) direct integration of Schrödinger equation [3.7] using pseudopotential technology [3.21].

The cluster and slab models, which is the subject of the present review, take a somewhat more specialized view of the surface. In the *cluster model* one assumes that a surface can be adequately represented by a small number of atoms. The electronic states of the cluster can be calculated by any of the traditional molecular orbital techniques [3.11]. The idea of regarding adsorption or chemisorption as essentially localized phenomenon is therefore quite attractive from the theoretical viewpoint. However, one must keep in mind that in practice the small particles which act as catalysts contain in excess of 100 metal atoms. It appears that a good account of one-electron eigenvalues of an adsorbate can be obtained by considering a modest-size cluster. However, extreme caution must be used in calculating the binding energy of an adsorbate using a similar-size cluster. The reason for this caution is that the positions of the substrate atoms are fixed by forces originating outside the cluster which are not incorporated in the calculation. In this sense a cluster is different from an ordinary stable molecule. A cluster can be made a more realistic representation of the surface by simply including more atoms. But one reaches the computational limit of the available computers before convergence in physical properties is achieved. Despite all these difficulties, cluster calculations have played a key role in elucidating electronic properties of chemisorption systems. Alternately, embedding techniques [3.22] can be used to study the effect of the rest of the substrate on the cluster eigenvalues. Cluster Bethe lattice [3.23] method can also be used for treating an "infinite" cluster incorporating only local coordination.

The *slab model* [3.24], on the other hand, has the complete two-dimensional periodicity, but the third dimension is terminated after a finite number of layers (usually 10-20). Many of the band structure techniques used in solid-state physics can be extended to treat this model, because the problem reduces to a band-structure calculation with a large unit cell. This model is quite suitable for treating ordered overlayers where interactions among adsorbate atoms are significant.

No exhaustive studies have been carried out to date which satisfactorily bridge the gap between the cluster and slab models. For example, it would be very interesting to know how the properties of a cluster lead into those of a macroscopic crystal as the cluster grows larger and larger. One knows that as more and more shells of neighboring atoms are added to a small central cluster the energy eigenvalues would become so dense that they would form continuous bands. In the limit of large number of atoms in the cluster, the energy "bands" of the cluster would be identical to those obtained using a periodic crystal. Thus an obvious, but difficult, question that requires an answer is that of how many atoms one would have to include in a cluster to get a density of states in substantial agreement with the energy-band picture? We shall attempt to answer this question for oxygen/aluminum system by presenting results for clusters of varying sizes and comparing these to energy bands obtained using slab models.

The organization of this chapter is as follows. In Sect.3.1 we give a general theoretical background and define the problem one is required to solve in electronic structure calculations. Section 3.1.1 discusses the Hartree-Fock approximation, and Sect.3.1.2 presents one simple form of the exchange potential ($x\alpha$). The implementation of the simplified exchange term for cluster models is discussed in Sects.3.1,3.4. Section 3.2 discusses details of the extended tight-binding method, the crystal potential, and matrix elements (Sect.3.2.1) with the slab model being described in Sect.3.2.2. Study of oxygen chemisorption on aluminum is presented in Sect.3.3. Four appendices have been given which describe various spectroscopies (Appendix 3.A), LCAO matrix elements (Appendix 3.B), Fourier transform of the exchange potential (Appendix 3.C) and embedding formalism (Appendix 3.D).

3.1 Theoretical Background

Electronic structure calculations usually solve the Schrödinger equation

$$H\Psi(\underline{x}) = E\Psi(\underline{x}) \quad , \tag{3.1}$$

where for an n electron system ($\underline{x}$ stands for both the spatial and spin coordinate),

$$\Psi(\underline{x}) = \Psi(\underline{x}_1,\underline{x}_2,\ldots,\underline{x}_n) \quad , \tag{3.2}$$

and the quantum-mechanical expression for the Hamiltonian in the *Born-Oppenheimer* approximation [3.25] is

$$H = -\sum_i \nabla_i^2 - 2\sum_i\sum_I \frac{Z_I}{|\underline{r}_i-\underline{R}_I|} + 2\sum_{i>j}\frac{1}{|\underline{r}_i-\underline{r}_j|} \quad , \tag{3.3}$$

and atomic (Rydberg) units with $\hbar = 1$, $e^2 = 2$, $m = 1/2$ are used throughout. Here Z_I is the atomic number of atom I. One can appreciate the complexity of calculating n-electron functions by noting that for a simple atom like $A\ell$, which has thirteen electrons, 10^{39} values are needed to specify Ψ at a coarse grid of 10 points per variable (excluding spin). Almost without exception then, one starts out with the so-called one-electron approximation where the n-electron wave function is expressed in terms of one-electron functions. Stated more precisely, the one-electron model is based on the assumption that an electron can be associate with a function $\phi_1(\underline{x}_1)$ (called spin orbital) which depends explicitly on the coordinates of only one electron. Formally, an n-electron wave function could be constructed as a simple product of n one-electron wave functions, but such a product does not satisfy the Pauli exclusion principle. Therefore, SLATER [3.26] proposed a determinantial form:

$$\Psi \approx \frac{1}{\sqrt{n!}} \begin{vmatrix} \phi_1(\underline{x}_1) & \phi_1(\underline{x}_2) & \cdots & \phi_1(\underline{x}_n) \\ \phi_2(\underline{x}_1) & \phi_2(\underline{x}_2) & \cdots & \phi_2(\underline{x}_n) \\ \cdot & \cdot & \cdots & \cdot \\ \phi_n(\underline{x}_1) & \phi_n(\underline{x}_2) & \cdots & \phi_n(\underline{x}_n) \end{vmatrix} \quad . \tag{3.4}$$

The above expression is often simplified by the following compact notation

$$\Psi \approx D[\phi_1(\underline{x}_1)\phi_2(\underline{x}_2) \ldots \phi_n(\underline{x}_n)] \quad . \tag{3.5}$$

In the approximated form, the n-electron wave function for aluminum atom can be specified by giving only 13×10^3 values in contrast to 10^{39} values noted earlier.

It should be realized that (3.4) or (3.5) is an approximation because the correct form of the wave function demands that the functional form for an orbital be such as to contain the coordinates of the n electrons, not of one only. In other words, we have assumed a functional form:

$$\phi_i = \phi_i(\underline{x}_i) \quad , \tag{3.6}$$

but we should really have

$$\phi_i = \phi_i(\underline{x}_i, \underline{x}_2, \ldots, \underline{x}_n) \quad . \tag{3.7}$$

To achieve this one can employ a linear combination of several determinants:

$$\Psi = \sum_i d_i D[\phi_1(\underline{x}_1), \phi_2(\underline{x}_2) \ldots \phi_n(\underline{x}_n)]_i \quad , \tag{3.8}$$

where the expansion coefficients d_i are determined by the use of variational principle. The expansion (3.8) is known as a configuration interaction wave function and is one way of introducing "correlation" [3.27].

3.1.1 Hartree-Fock Approximation

If one uses a single-determinant wave function and ensures that the energy corresponding to the function of (3.4) is as low as possible (variational principle) then the spin orbitals entering the Slater determinant satisfy an equation known as the Hartree-Fock equation, namely,

$$\left[-\nabla_1^2 - \sum_I \frac{2Z_I}{|\underline{r}_1 - \underline{R}_I|} + \sum_\ell (\hat{J}_\ell - \hat{K}_\ell)\right] \phi_i(\underline{r}_1) = \varepsilon_i \phi_i(\underline{r}_1) \quad . \tag{3.9}$$

The operation of the "Coulomb" $\hat{J}_\ell$ and "exchange" $\hat{K}_\ell$ operators on the arbitrary function ϕ_i is defined by the relations

$$\hat{J}_\ell \phi_i(\underline{r}_1) = 2 \int \frac{|\phi_\ell(\underline{r}_2)|^2}{r_{12}} d\underline{r}_2 \phi_i(\underline{r}_1) \tag{3.10}$$

$$\hat{K}_\ell \phi_i(\underline{r}_1) = 2 \int \frac{\phi_\ell^*(\underline{r}_2)\phi_i(\underline{r}_2)}{r_{12}} d\underline{r}_2 \, \phi_\ell(\underline{r}_1) \quad . \tag{3.11}$$

Note that the term $\ell = i$ in the exchange term is the usual self-interaction of the electron in orbital ϕ_i, namely,

$$-2 \int \frac{|\phi_i(\underline{r}_2)|^2}{r_{12}} d\underline{r}_2 \phi_i(\underline{r}_1) \quad .$$

Thus the Hartree-Fock exchange term includes both the self-interaction and the terms for which $\ell \neq i$. Sometimes only these later terms, for $\ell \neq i$, are referred to as the exchange terms. However, when we refer to the exchange term we will assume that the self-interaction is included in it. It should be noted that in the exchange term only nonvanishing contributions come from the spin orbitals ϕ_ℓ which have the same spin σ^i as ϕ_i.

The orbital energies ε_i can be related to the ionization energies of the system. From (3.9) we can write

$$\varepsilon_i = h_{i,i} + \sum_\ell J_{\ell,i} - \sum_\ell K_{\ell,i} \delta(\sigma^\ell, \sigma^i) \quad , \tag{3.12}$$

where

$$h_{i,i} = \int \phi_i^*(\underline{r}_1) \left[-\nabla_1^2 - \sum_I \frac{2Z_I}{|\underline{r}_1 - \underline{R}_I|} \right] \phi_i(\underline{r}_1) d\underline{r}_1$$

$$J_{\ell,i} = \int \phi_i^*(\underline{r}_1) \hat{J}_\ell \phi_i(\underline{r}_1) d\underline{r}_1 \left(= \int \phi_\ell^*(\underline{r}_1) \hat{J}_i \phi_\ell(\underline{r}_1) d\underline{r}_1 \right)$$

$$= 2 \iint \frac{\phi_i^*(\underline{r}_1)\phi_i(\underline{r}_1)\phi_\ell^*(\underline{r}_2)\phi_\ell(\underline{r}_2)}{r_{12}} d\underline{r}_1 d\underline{r}_2 \tag{3.13}$$

$$K_{\ell,i} = \int \phi_i^*(\underline{r}_i) \hat{K}_\ell \phi_i(\underline{r}_1) d\underline{r}_1 \left(= \int \phi_\ell^*(\underline{r}_1) \hat{K}_i \phi_\ell(\underline{r}_1) d\underline{r}_1 \right)$$

$$= \iint \frac{\phi_i^*(\underline{r}_1)\phi_\ell(r_1)\phi_i(\underline{r}_2)\phi_\ell^*(\underline{r}_2)}{r_{12}} d\underline{r}_1 d\underline{r}_2 \quad . \tag{3.14}$$

Incidentally, one can obviously see that $J_{i,i} = K_{i,i}$. In the special case of a closed-shell system where each orbital is doubly occupied, one may write

$$\varepsilon_i = h_{i,i} + \sum_\ell (2J_{\ell,i} - K_{\ell,i}) \quad . \tag{3.15}$$

The expression for the Hartree-Fock total energy is

$$E = 2\sum_i h_{i,i} + \sum_i \sum_\ell (2J_{\ell,i} - K_{\ell,i}) \quad . \tag{3.16}$$

If one computed the total energy of the ionized system in which the ith spin orbital is missing but all the remaining orbitals are same before and after the ionization and subtracted that from (3.16), the energy difference is precisely the eigenvalue ε_i given by (3.15). Thus, the physical interpretation is that it costs $-\varepsilon_i$ to remove an electron from the orbital ϕ_i provided the orbitals are not permitted to relax. This is a statement of KOOPMANS' theorem [3.28]. However, deviations from Koopmans' theorem are to be expected since electrons do relax after an excitation. The energy of the ionized system will be somewhat lower when the orbitals are allowed to be modified as compared to the values appropriate for the neutral system (otherwise the orbitals would not be relaxing in the first place). Thus a more accurate estimate of the ionization energy is obtained by solving separate self-consistent problems for the neutral and for the ionized system and then subtracting the energies of these two problems. It is clear that Koopmans' theorem gives too high a value for the ionization energy.

3.1.2 Statistical Exchange Approximation

The nonlocal exchange term in the Hartree-Fock equation (3.9) is different for each spin orbital and is therefore quite complicated to manage, especially for polyatomic systems. Therefore, SLATER [3.29] proposed the use of the xα approximation, which amounts to replacing the exchange term in (3.9) as follows:

$$\sum_\ell \hat{K}_\ell \rightarrow 6\alpha \left(\frac{3}{4\pi}\rho_\uparrow\right)^{1/3} \quad , \tag{3.17}$$

where $\rho\uparrow$ is the spin-up part of the charge density ($\rho_\uparrow \approx \rho_\downarrow \approx \rho/2$). The one-electron equation in the xα approximation is

$$\left[-\nabla_1^2 - \sum_I \frac{2Z_I}{|r - R_I|} + \sum_\ell \hat{J}_\ell + V_{x\alpha}(1)\right] \phi_i(\underline{r}_1) = \varepsilon_i^\alpha \phi(\underline{r}_1) \quad , \tag{3.18}$$

where

$$V_{x\alpha} = -6\alpha \left[\frac{3}{8\pi}\rho\right]^{1/3} \quad . \tag{3.19}$$

The original derivation due to SLATER [3.29] lead to $\alpha = 1$, and the alternative derivation of GASPER, KOHN, and SHAM [3.30] lead to $\alpha = 2/3$. More generally, α is treated as a parameter which can be determined from atomic calculations (xα approximation) [3.31]. The eigenvalues are

$$\varepsilon_i^\alpha = h_{i,i} + \sum_\ell n_\ell J_{\ell,i} + \int \rho(\underline{r}) V_{X\alpha}(\underline{r}) d\underline{r} \quad . \tag{3.20}$$

The total energy expression in the xα approximation is given by

$$\langle E_{X\alpha}\rangle = \sum_i n_i h_{i,i} + \frac{1}{2} \sum_{i,j} n_i n_j J_{i,j} + \frac{1}{2} \int \rho(\underline{r}) U_{X\alpha}(\underline{r}) d\underline{r} \quad , \tag{3.21}$$

where

$$U_{X\alpha}(\underline{r}) = \frac{3}{2} V_{X\alpha}(\underline{r}) \quad , \tag{3.22}$$

and the occupation number n_i may take on fractional values. It can be easily shown that the eigenvalues ε_i^α are given by [3.32]

$$\varepsilon_i^\alpha = \frac{\partial}{\partial n_i} \langle E_{X\alpha}\rangle \quad . \tag{3.23}$$

Thus we would expect ε_i^α to be quite different from the Hartree-Fock eigenvalues. Recall that the Hartree-Fock eigenvalues are obtained as a finite difference of energies for two states for which the occupation number n_i of the ith spin orbital differs by unity, whereas xα eigenvalues are obtained from a partial derivative.

It should be pointed out that ε_i^α are not approximations to ionization potentials. To obtain ionization and excitation energies, one must use the SLATER transition method [3.32], which consists of solving a self-consistent field problem, not for the ground state, but for a state in which the occupation numbers of the various spin orbitals are halfway between those of the initial and final states. The difference in eigenvalues for such a transition state calculation can be identified with the excitation energy. For an ionization process the transition state is one in which the electron being removed is located half in the state from which it is coming and half in the state at infinite distance from the nucleus to which it is going. This latter half electron has no effect on the problem. The ionization energy is simply equal to $-\varepsilon_i^\alpha$ of the state from which half an electron has been removed. The transition state procedure takes care of electronic relaxation effects inherent in the excitation process, a fact which is disregarded in the conventional use of the Hartree-Fock method and Koopmans' theorem for finding excitation energies.

3.1.3 SCF-xα-SW Method

One powerful method for solving the xα equations (3.18) is based on the scattered wave (SW) formalism of SLATER and JOHNSON [3.33]. The self-consistent-field xα-scattered-wave method [3.33] divides a molecule (or cluster) into three fundamental types of regions: (1) atomic, (2) interatomic, and (3) extraatomic. Wave functions and potentials are treated differently in these regions. In the atomic region, the wave function is expanded in spherical harmonics times a radial function to be determined by numerical integration, and the potential is considered to be spherically symmetric. A similar expansion is performed for the extraatomic region. In the interatomic region, the potential is assumed to be constant and the solutions are written as a sum of spherical harmonics times spherical Bessel functions of imaginary arguments (for $\varepsilon_i^\alpha < 0$). Imposing the requirement that the interatomic wave functions and their first derivatives should be continuous with the solutions to the atomic and extraatomic regions leads to a set of secular equations in which the energy occurs as a parameter. It is therefore indirectly determined by the consistency of the linear equations and is found by interpolation. The scattered wave formalism is a convenient means of setting up the secular equation. The expansion used in the atomic and interatomic regions are analogous to the KORRINGA-KOHN-ROSTOKER (KKR) method [3.34] of band theory. The choice of potential described above usually goes by the name of the "muffin-tin" potential. For those molecular orbitals which have a significant fraction of their charge density localized in the interatomic region, the muffin-tin potential leads to less accurate eigenvalues. These errors can be reduced by use of perturbation theory corrections [3.35] and by an overlapping atomic spheres scheme [3.36].

3.1.4 LCAO-Xα Method

In this method individual one-electron wave functions $\phi_i(\underline{r})$ of the polyatomic system are approximated by a linear combination of atomic orbitals (LCAO), $\chi_m(\underline{r})$, centered on various atomic positions,

$$\phi_i(\underline{r}) = \sum_m C_{mi}\chi_m(\underline{r}) \quad , \tag{3.24}$$

The atomic orbitals themselves are taken to be a linear combination of basis functions, usually exponentials $\sim r^n e^{-\alpha r}$ or Gaussians $\sim r^n e^{-\alpha r^2}$. Numerical basis sets obtained from solutions of the free atom problem are also frequently used [3.18b]. The eigenvalues and expansion coefficients are obtained by solving the eigenvalue problem,

$$\underset{\approx}{H}\underset{\sim}{C}_i = \varepsilon_i \underset{\approx}{S}\underset{\sim}{C}_i \quad , \tag{3.25}$$

where $\underset{\approx}{H}$ and $\underset{\approx}{S}$ are the Hamiltonian and overlap matrices, respectively, whose elements are given by

$$H_{mn} = \int \chi_m^* \hat{H} \chi_n d\tau \quad , \tag{3.26}$$

$$S_{mn} = \int \chi_m^* \chi_n d\tau \quad . \tag{3.27}$$

In ab initio calculations, the multicenter integrals (3.26-27) are evaluated explicitly for a given set of χ_m and $\hat{H}$. The discrete variational method [3.37] is commonly used to compute the necessary matrix elements. The intractability of a direct evaluation of the matrix elements of the exchange potential (being proportional to the 1/3 power of the charge density) has resulted in various fitting schemes for representing the xα potential in terms of analytical functions. This has led to a large number of significantly different LCAO-xα methods [3.18].

A large number of semiempirical developments [3.17], where one parameterizes or assumes some simple forms for the matrix elements, have also taken place. One often encounters the Coulomb and resonance integrals defined, respectively, as

$$\alpha_i \equiv H_{ii} = \langle \chi_i|\hat{H}|\chi_i\rangle \quad , \tag{3.38}$$

$$\beta_{ij} \equiv H_{ij} = \langle \chi_i|\hat{H}|\chi_j\rangle \quad . \tag{3.29}$$

The Coulomb integral α_i is approximately equal to the atomic term value to which the function χ_i belongs. The resonance integrals describe the interaction of different atomic orbitals and are often taken proportional to overlap integrals S_{ij}. Semiempirical methods [3.38] are widely used since the empirical values of the parameters take implicit account of a number of factors omitted in the nonempirical approach.

3.2 ETB-xα Method

The extended tight-binding (ETB) method is an application of the one-electron model to crystals (band theory). It is essentially identical to the molecular-orbital LCAO method of quantum chemistry. The study of electronic structure of a crystal reduces within the framework of band theory to the solution of

$$[-\nabla^2+V(\underline{r})]\phi_n(\underline{k},\underline{r}) = \varepsilon_n(\underline{k})\phi_n(\underline{k},\underline{r}) \quad , \tag{3.30}$$

where $V(\underline{r})$ is an effective potential of the crystal which for a periodic lattice characterized by a set of lattice vectors $\underline{R}_\nu$, satisfies the condition

$$V(\underline{r}) = V(\underline{r}+\underline{R}_\nu) \quad . \tag{3.31}$$

Also the translational symmetry allows one to introduce the wave vector $\underline{k}$ and the band index n to characterize each solution. Equation (3.31) also implies that the solution of Eq. (3.30) should satisfy a boundary condition,

$$\phi(\underline{r}+\underline{R}_\nu) = \exp(i\underline{k}\cdot\underline{R}_\nu)\phi(\underline{r}) \quad , \tag{3.32}$$

which is referred to as the Bloch's theorem [3.39]. To be consistent with this condition, one chooses the following LCAO expansion:

$$\phi_n(\underline{k},\underline{r}) = \frac{1}{\sqrt{N}} \sum_\nu \sum_\mu \exp(i\underline{k}\cdot\underline{R}_\nu)C_{\mu n}(\underline{k})\chi_\mu(\underline{r}-\underline{R}_\nu) \quad , \tag{3.33}$$

where $\chi_\mu(\underline{r})$ ($\mu=1,2,\ldots M$) are the atomic orbitals in the reference unit cell, $C_{\mu n}(\underline{k})$ are the expansion coefficients, and the summation over lattice vectors $\underline{R}_\nu$ extends over all the lattice points N in the crystal. Sometimes, one also defines the Bloch sums as

$$X_\mu(\underline{k},\underline{r}) = \frac{1}{\sqrt{N}} \sum_\nu \exp(i\underline{k}\cdot\underline{R}_\nu)\chi_\mu(\underline{r}-\underline{R}_\nu) \quad , \tag{3.34}$$

then the expansion (3.33) takes more characteristic LCAO form, namely,

$$\phi_n(\underline{k},\underline{r}) = \sum_\mu C_{\mu n}(\underline{k}) X_\mu(\underline{k},\underline{r}) \quad . \tag{3.35}$$

The expansion coefficients in the above equation are obtained using the variational principle,

$$\frac{\partial}{\partial C_\mu(\underline{k})} \left(\frac{<\phi|H|\phi>}{<\phi|\phi>} \right) = 0 \quad , \quad \mu = 1,2,\ldots M \quad . \tag{3.36}$$

This leads to the eigenvalue problem

$$\sum_{\mu'} \left[H_{\mu,\mu'}(\underline{k}) - \varepsilon_n(\underline{k}) S_{\mu,\mu'}(\underline{k}) \right] C_{\mu n}(\underline{k}) = 0 \quad , \tag{3.37}$$

where the $\underline{k}$-dependent Hamiltonian and overlap matrices are

$$H_{\mu,\mu'}(\underline{k}) = \sum_\beta \exp(i\underline{k}\cdot\underline{R}_\beta) h_{\mu,\mu'}(\underline{R}_\beta) \quad , \tag{3.38}$$

$$S_{\mu,\mu'}(\underline{k}) = \sum_\beta \exp(i\underline{k}\cdot\underline{R}_\beta) s_{\mu,\mu'}(\underline{R}_\beta) \quad . \tag{3.39}$$

Equations (3.38,39) have been obtained using the translational symmetry of the system. The elements of the matrices $h(\underline{R}_\beta)$ and $s(\underline{R}_\beta)$ are defined as follows:

$$h_{\mu,\mu'}(\underline{R}_\beta) = \int \chi_\mu^*(\underline{r}) \hat{H} \chi_{\mu'}(r-\underline{R}_\beta) d\underline{r} \quad , \tag{3.40}$$

$$s_{\mu,\mu'}(\underline{R}_\beta) = \int \chi_\mu^*(\underline{r}) \chi_{\mu'}(r-\underline{R}_\beta) d\underline{r} \quad . \tag{3.41}$$

Because the summation in (3.38) extends over the entire direct lattice, thus requiring a large number of integrals, it is quite time consuming to evaluate all the potential matrix elements of (3.40). In general these lattice sums are not rapidly convergent. Thus in earlier applications of tight binding methods drastic approximations were introduced. For example, these multicenter integrals were usually neglected altogether or else were parameterized. SLATER and KOSTER [3.40] suggested using the LCAO method as an interpolation scheme. They noted that accurate solutions at high symmetry points in the Brillouin zone could be obtained more simply using other techniques. One could then treat $h_{\mu,\mu'}(\underline{R}_\beta)$ as parameters for fitting the calculated energy bands at such points and use these parameters to calculate eigenvalues at arbitrary $\underline{k}$ points.

Recent renewed interest in the tight-binding method as a first-principles method has come about because of tremendous advances made in quantum chemistry [3.41] using the Gaussian-type orbital (GTO) and the advent of powerful computers. It has been found that if for the radial part of χ_μ one chooses a function of the form $\sim r^n \exp(-\alpha r^2)$, then all the necessary matrix elements can be reduced to analytic forms [3.42]. For solids the potential matrix elements can be obtained in a closed form using a Fourier representation of the crystal potential:

$$V(\underline{r}) = \sum_{\underline{G}} V(\underline{G}) \exp(-i\underline{G}\cdot\underline{r}) \quad , \tag{3.42}$$

where $\underline{G}$ are the reciprocal lattice vectors and $V(\underline{G})$ are the Fourier coefficients. This brings in a sum over reciprocal lattice vectors and this sum may be slowly convergent. Thus it is necessary that the matrix elements of the type $\langle\chi_\mu|\exp(-i\underline{G}\cdot\underline{r})|\chi_{\mu'}\rangle$ should be calculated rapidly. GTO's do indeed fulfill this requirement.

A fundamental question for solids, however, is the choice of the crystal potential itself. To large extent, the representation used for this potential would determine the speed and accuracy with which the multicenter integrals can be evaluated. Therefore, let us turn to the crystal potential to be used in the ETB calculations. Having given a specific algebraic form for the crystal potential, we would then be able to give explicit expressions for its matrix elements.

3.2.1 Crystal Potential and Matrix Elements

The simplest kind of approximation for the crystal potential $V_{cry}(\underline{r})$, is one where neutral atom potentials are superimposed to generate $V_{cry}(\underline{r})$ (overlapping atomic potential approximation). Thus

$$V_{cry}(\underline{r}) = \sum_{\nu,\mu} V_a(r_{\nu\mu}) \quad , \tag{3.43}$$

where $r_{\nu\mu} = |\underline{r}-\underline{R}_\nu-\underline{\tau}_\mu|$ and (the spherically symmetric)

$$V_a(r) = -\frac{2Z}{r} + 2\int \frac{\rho(\underline{r}')}{|\underline{r}-\underline{r}'|}\, d\underline{r}' - 3\alpha\left(\frac{3}{\pi}\right)^{1/3} \rho^{1/3}(r) \quad , \tag{3.44}$$

the last term being the exchange potential in the statistical exchange ($x\alpha$) theory [3.32]. If we attempted to expand $V_{cry}(\underline{r})$ in a Fourier representation according to (3.42), a sum over a large number of reciprocal lattice vectors has to be performed. The slow convergence is caused by the presence of the $-2Z/r$ term about each nucleus. Thus in practice one uses an Ewald-type procedure [3.43] where the crystal potential is split into two parts:

$$V_{cry}(\underline{r}) = \sum_{\nu,\mu} \{V_c(r_{\nu\mu}) + [V_a(r_{\nu\mu}) - V_c(r_{\nu\mu})]\} \quad , \tag{3.45}$$

where [3.44]

$$V_c(r) = \frac{C_1}{r} e^{-a_1 r^2} + \frac{C_2}{r} e^{-a_2 r^2} + \sum_{i=3}^{m} C_i \exp(-a_i r^2) \tag{3.46}$$

behaves like $-2Z/r$ for small values r if $C_1+C_2=-2Z$. Various coefficients C_i and exponents a_i are obtained by nonlinear least square fitting to (3.44). The remaining part $[V_a-V_c]$ is now a relatively smooth function of $\underline{r}$ and can be expanded in a rapidly convergent Fourier series:

$$V_a(\underline{r}) = V_c(\underline{r}) \equiv V_S(\underline{r}) = \sum_{\underline{G}} V_S(\underline{G}) e^{-i\underline{G}\cdot\underline{r}} \quad . \tag{3.47}$$

Thus the final expression for the crystal potential becomes

$$V_{cry}(\underline{r}) = \sum_{\nu,\mu} \left[\frac{C_1(\mu)}{r_{\nu\mu}} \exp(-a_1(\mu) r_{\nu\mu}^2) + \frac{C_2(\mu)}{r_{\nu\mu}} \exp(-a_2(\mu) r_{\nu\mu}^2) \right. \tag{3.48}$$

$$\left. + \sum_{i=3}^{m} C_i(\mu) \exp(-a_i(\mu) r_{\nu\mu}^2) + \sum_{\underline{G}} V_S(\underline{G}) \exp(-i\underline{G}\cdot\underline{r}_{\nu\mu}) \right] \quad ,$$

where $C_1(\mu)+C_2(\mu)=-2Z_\mu$ and Z_μ is the atomic number of the μth atom.

It should be observed that if a large number of terms are used in (3.46), then the contribution to matrix elements from the smooth part, $V_S(\underline{r})$, may be small, In that case one needs only a few reciprocal lattice vectors. On the other hand, if only a few terms are used in (3.46), one must span a larger portion of the reciprocal space. In practice one chooses an optimum value depending on the system under consideration.

We now define the GTO basis set. A general Gaussian function (α) centered on site $\underline{A}$ is defined as [3.45-46]

$$(\alpha) \equiv (\underline{A},\alpha,\ell,m,m) = N(x-A_x)^\ell (y-A_y)^m (z-A_z)^n e^{-\alpha r_A^2} \quad , \tag{3.49}$$

where N is the normalization constant. The type of the Gaussian function is determined by integers (ℓ,m,n). For example, an s-type orbital has $\ell=m=n=0$. The orbital exponents α for various atoms have been extensively tabulated in the literature [3.46]. In our calculations, matrix elements of various operators would be required between primitive Gaussians of the type (3.49). Apart from overlap and kinetic energy matrix elements, potential matrix elements require knowledge of integrals of the type $\langle(\alpha)_1|[\exp(-\beta r^2)]/r|(\alpha)_2\rangle$, $\langle(\alpha)_1|\exp(-\beta r^2)|(\alpha)_2\rangle$, and $\langle(\alpha)_1|\exp(-i\underline{G}\cdot\underline{r})|(\alpha)_2\rangle$. These matrix elements have been evaluated in Appendix 3.B using the work by TAKETA et al. [3.47]. It suffices here to mention that all these expressions have been carefully programmed for a systematic evaluation on a digital computer.

Before closing this subsection we should point out that a better approximation for generating the crystal potential is the so-called overlapping charge density approximation, since the exchange potential is a nonlinear ($\rho_T^{1/3}(\underline{r})$) function of the electronic charge density. This would require a different technique for obtaining Fourier coefficients in (3.47). A complete discussion for obtaining Fourier coefficients of the exchange potential is given in Appendix 3.C. Once Fourier coeffiecients are obtained, the rest of the treatment is same as in the overlapping atomic potential case.

3.2.2 ETB-Slab Model

The introduction of a surface results in a loss of translational invariance in the direction perpendicular to the surface. Thus the potential is no longer three-dimensional periodic and (3.31) is not valid. This also implies that we cannot express the crystal potential in a Fourier representation according to (3.42). However, such an expansion was a key feature in making the troublesome three-center integrals convergent. At this point we can proceed in one of two ways.

We can "artificially" generate [3.18-24] three-dimensional periodicity by choosing a somewhat unusual unit cell (the slab model). To represent a surface we choose a slab whose base is the two-dimensional unit cell and the third dimension contains a finite number of atomic layers. The periodicity along the third dimension can be generated by putting many slabs on top of each other with empty space interposed between any two slabs. If the empty space has sufficient thickness, the interaction between two consecutive slabs is negligible and one has produced a free surface with a somewhat artificial three-dimensional periodicity. The repeat distance along the third dimension is equal to the thickness of the slab plus the width of empty space (L) between two slabs. It should be noted that one cannot arbitrarily increase L because shells of reciprocal lattice vectors would lie too close to each other. It would then be difficult to span any sensible portion of the reciprocal space without including a large number of $\underline{G}$ vectors in the summation of (3.42).

Another way of implementing the slab model is to avoid using the Fourier representation altogether. In this scheme one constructs a single slab of several atomic layers and uses the representation (3.46) for the crystal potential. If a sufficient number of terms are indicated in (3.46), then the crystal potential can indeed be adequately represented. This would require numerically small exponents in the expansion (3.46) to represent the long-range potential, and hence out of necessity, three-center integrals would have to be calculated. As discussed earlier (Appendix 3.B) the matrix elements can once again be evaluated analytically. However, there is a price to be paid. Since some of the exponents in (3.46) would be quite small, lattice summations in the matrix elements evaluation have to be carried out to far distant neighbors before convergence is achieved. Both schemes have been used in the literature [3.18,24] with considerable success.

3.3 Oxygen Chemisorption on Aluminum

Chemisorption is defined as the adsorption of a single layer of chemically bound atoms or molecules on the surface of a solid. Chemisorption involves forces similar to those occurring in chemical reactions. A gas will chemisorb under conditions where one might expect at least incipient chemical reactions between the gas and the surface. Thus one would expect oxygen to chemisorb on various surfaces of aluminum because of the known existence of aluminum oxides. Chemisorption is also of interest to those engaged in the study of catalysis. It is related to catalysis in the sense that at least one of the reactants must be chemisorbed before a catalytic reaction will take place. Our emphasis is directed toward the manner in which our knowledge relative to the electronic structure of the chemisorbed state may be important for learning the electronic properties of solid surfaces.

There is considerable theoretical and experimental interest in the chemisorption of oxygen on aluminum because it is a good prototype system. Theoretically, aluminum is easy to handle because there are no d bands to complicate the calculations. Consequently, a number of theoretical methods such as adatom jellium [3.48-49], SCF-xα cluster [3.50-55], ETB-xα slab [3.56-57], and others [3.58-60] have been employed for electronic structure calculations of the oxygen-aluminum system. Energy band structure calculational results for Al(001), Al(110), and Al(111) surfaces are also available [3.61]. On the experimental side there is an equally intense activity. Many experimental techniques such as quartz crystal microbalance for mass-gain determination [3.62], Auger electron spectroscopy [3.63-65], work function measurements [3.64,66], secondary-ion mass spectroscopy [3.67], X-ray photoemission measurement of the angular dependence of various plasmon-satellite peak intensities [3.68-69], partial protoyield spectroscopy [3.70], XPS [3.71], ultraviolet and synchrotron radiation photoemission spectroscopy [3.49,72-80], LEED [3.81-83], SEXAFS [3.84], EAPFS [3.85], and Ellipsometry [3.86] have been employed. Oxidation of aluminum at atmospheric pressure has also been studied extensively [3.87-96]. More recently, angle-resolved ultraviolet photoemission spectroscopy (ARUPS) has been used [3.97-99] to study the interaction among chemisorbed oxygen atoms on aluminum surface.

Our interest lies in determining the geometric location of oxygen atoms and the resultant electronic structure. Therefore in what follows, we shall calculate the electronic structure for an assumed reasonable geometry and compare the results with the relevant experimental data.

3.3.1 Cluster Approach

One of the earlier cluster calculation for O-Aℓ is due to HARRIS and PAINTER [3.50]. They calculated total energy of the cluster $O(A\ell)_5$ within the SCF-xα muffin-tin approximation to determine the equilibrium position of oxygen on Aℓ(001). No minimum in the total energy was found due perhaps to the failure of the muffin-tin approximation or to the small cluster size as discussed earlier. Subsequently extensive cluster model calculations based on the SCF-$X\alpha$-SW method have been performed by MESSMER and coworkers [3.51-54] for the three low index faces. Let us discuss in detail the cluster results by MESSMER and SALAHUB (MS) for the O+$A\ell$(001) system [3.51].

The Al(001) surface has its atoms arranged in a square network of side $a/\sqrt{2}$ (=2.86 Å), the lattice constant a=4.04 Å. The distance between successive layers of atoms along the z direction is a/2. MS used clusters consisting of 5, 9, and 25 aluminum atoms in interaction with 1, 4, or 5 oxygen atoms at various positions above, in, and below the surface. The largest cluster (25 aluminum atoms) has 12 aluminum atoms in the first plane, 9 in the second, and 4 in the third and is schematically shown in Fig.3.1. An oxygen atom incorporated in the top layer (z=0.0) in the fourfold site does not strain any of the bond lengths because

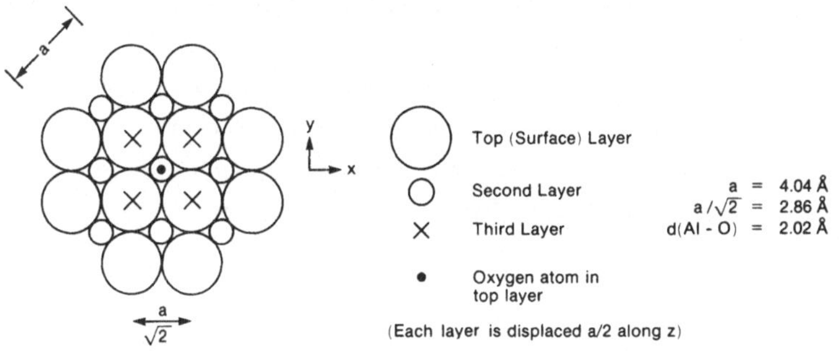

Fig. 3.1. Schematic view of the 25-atom cluster for O-Al(001).

d($A\ell$-O)=2.02 Å is still larger than the $A\ell$-O bond length of 1.97 Å in $A\ell_2O_3$. Thus there is no steric hinderance and oxygen may well penetrate the (001) surface.

From a cluster model calculation one usually gets a discrete set of orbital energies, but experimentally one measures the "density of states", a quantity which is calculated for bulk solids in band theory. To facilitate a comparison of the calculated results with the results of photoemission experiments, it is convenient to define the total and local density of states (TDOS, LDOS). The total density of states, D(E), are obtained by assigning to each computed molecular energy ε_i, a Gaussian (of full width at half maximum 2.335Γ)

$$D(E) = \frac{1}{\Gamma(2\pi)^{\frac{1}{2}}} \sum_i N_i \exp[-(E-\varepsilon_i)^2/2\Gamma^2] \ , \qquad (3.50)$$

where N_i is the occupancy of the eigenstate ε_i. To calculate the local density of states (LDOS) at site j,

$$D_j(E) = \frac{1}{\Gamma(2\pi)^{\frac{1}{2}}} \sum_i N_i f_i(j) \exp[-(E-\varepsilon_i)^2/2\Gamma^2] \ . \tag{3.51}$$

Here the local weighting factor is given by total charge on a particular atomic center j,

$$f_i(j) = Q_i(j)/\sum_k Q_i(k) \quad , \tag{3.52}$$

where the k summation extends over all atoms in the cluster. The $Q_i(j)$'s are calculated self-consistently in the cluster calculations being discussed here. The solid-state analogs of (3.50-52), which we will use later, are defined in [3.100].

Local density of states for the oxygen atom in Al_5+O, Al_9+O, Al_{25}+O clusters calculated by MS are shown in Fig.3.2. It is clear that both the Al_5 and Al_9 clusters show two oxygen-related peaks, whereas the Al_{25}+O cluster shows a third peak which is qualitatively different from the two smaller clusters. MS concluded that the 25-atom cluster had the possibility of representing the Al(001) surface and went on to explore it further. Figure 3.3a shows the partial density of states for an oxygen atom in an Al_{25}+O cluster for the case where the oxygen atom is incorporated in the top surface plane (Z_0=0.0). The TDOS curve has been further de-

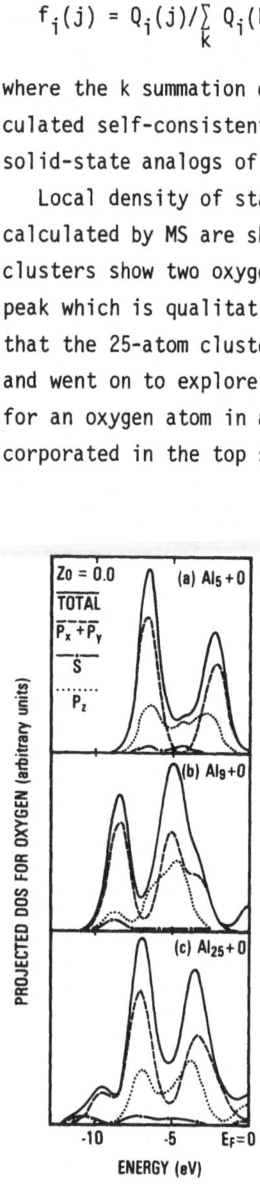

Fig. 3.2. Local density of states for the oxygen atoms vs. orbital energy (eV relative to Fermi level E_F) for aluminum clusters interacting with an oxygen atom in the first surface plane (Z_0=0 a.u.) (from [3.51]).

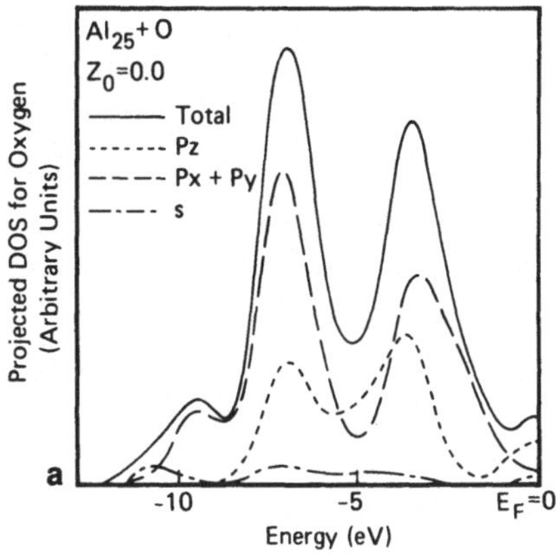

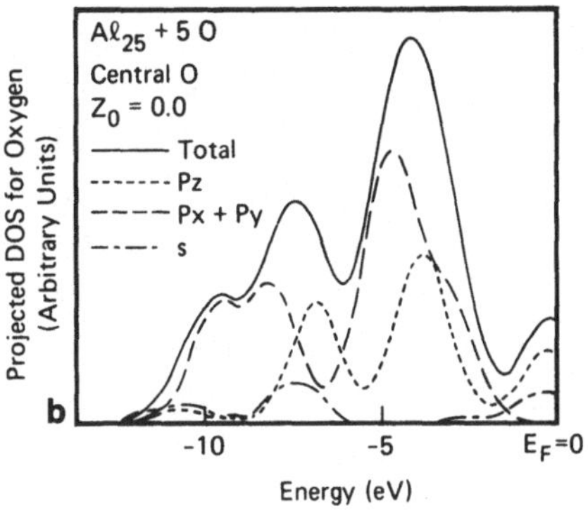

Fig. 3.3(a) Local density of states for the oxygen atom for Al_{25}+O, Z_0=0 a.u. (b) local density of states for the central oxygen atom for Al_{25}+5O, Z_0=0 a.u. (from [3.51])

composed into contributions from s functions, p_z functions, and p_x+p_y functions. For the model in which oxygen atoms are situated in a plane of aluminum atoms, LDOS show three energy regions of significant oxygen character about -9.5, -7, and -3 eV relative to the Fermi energy. Inspection of the wave functions for the corresponding levels show that the -9.5 eV peak is due mainly to binding combinations of in-plane oxygen $p_{x,y}$ orbitals with aluminum s and p orbitals. The -7 eV peak arises from both in-plane ($p_{x,y}$) and out-of-plane oxygen p_z orbitals again in bonding combinations. The orbitals responsible for the -3 eV peak were classified as nonbonding with respect to aluminum-oxygen interactions. The first two peaks (-9.5 and -7 eV) have been observed in many photoemission studies [3.49,72-79], but there is no general agreement on the origin of the third peak. EBERHARDT and KUNZ [3.77], using 32-eV photons, observed a peak at -2 eV for O-Al(001) up to 30 Langmuir exposures. They state that "this extra structure may no longer be resolved after larger exposures because the tail of the normal O(2p) resonance extends up to the Fermi level." Others have interpreted this structure due to the surface state of clean Al. Furthermore, as will be seen below, this structure has not been seen on the Al(111) surface. In another calculation, MS placed the oxygen atom at Z=1.06 Å and found two oxygen related peaks at about -2.5 eV and -5 eV. This, being in total disagreement with all the available experimental data, led to the conclusion that the oxygen atoms are not at this position.

MS also studied the effect of adsorbate-adsorbate interaction on the density of states by using an Al_{25}+5O cluster. The calculation was carried out for the case of five oxygen atoms in the five fourfold sites of the first surface layer. Since the central site has an environment closer to that for an infinite surface, the LDOS shown in Fig.3.3b were obtained for the central oxygen. On comparing Figs.3.3a,b one notices that the three peaks structure stays reasonably intact but with somewhat different widths and intensities. An obvious important difference is that the -3 eV peak has gained intensity at the expense of the -7 eV peak. By analyzing the orbital density of states one can draw the following inference about the effect of lateral interactions. In the absence of lateral interactions the in-plane ($p_{x,y}$) and the out-of-plane (p_z) components of the -7 and the -3 eV peaks have their maxima at almost the same energies. The major effect of adding on the four nearby oxygen atoms (or turning on the lateral interactions) is to produce a relative shift between the in-plane and out-of-plane components by about 1 eV. The $p_{x,y}$ components have moved further away from the Fermi energy while p_z components have moved towards it. We will see below in the slab calculation that this trend is very well reproduced when complete adlayer interactions are taken into account. The results for the O-Al(001) can be summarized by stating that oxygen-related features appear at -9.5, -7, and -3 eV below E_F for the case of oxygen incorporation in the first surface layer. There is sufficient room for an oxygen atom to fit in the fourfold site on Al(001) surface leading to a bond length of d(Al-O)=2.02 Å. This is quite a reasonable value because for bulk Al_2O_3, d(Al-O)=1.97 Å.

The incorporation in the Al(111) surface may not be too favorable because of an unreasonably small Al-O bond length (1.65 Å) that is considerably less than the sum of the atomic radii of oxygen and aluminum being equal to 1.85 Å (R_O=0.6 Å, R_{Al}=1.25 Å). Figure 3.4 shows the geometrical arrangement of the (001) and (111) surfaces side by side. The interlayer spacing for the (001) surface is 2.02 Å as opposed to 2.34 Å for Al(111). Thus one might expect that the electronic structure of oxygen chemisorbed on Al(111) surface to be quite different, and this we briefly take up now following the work of SALAHUB, ROCHE, and MESSMER (SRM) [3.52]. There is good evidence [3.79,82-83] that oxygen forms an ordered overlayer on Al(111).

On Al(111) there are two distinct threefold sites for oxygen chemisorption (Fig.3.5). One can place the oxygen atom above the threefold hollow or the threefold blocked site. The former site has no atom on the z axis in the second layer (an "FCC" site) while the latter does (an "HCP" site). Both of these distinct sites, denoted by A and B, respectively, have been considered by SRM. They concluded that

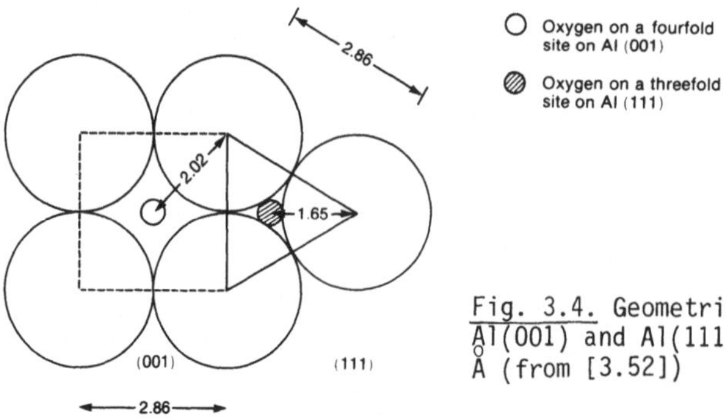

Fig. 3.4. Geometrical arrangement of the atoms on Al(001) and Al(111) surfaces. All distances are in Å (from [3.52])

$Al_{19}^{(A)}$ and $Al_{22}^{(B)}$ are likely to be reasonably converged models of the (111) surface. The major evidence used for this conclusion was that the calculated bandwidth for $Al_{19}^{(A)}$ (9.6 ev) and $Al_{22}^{(B)}$ (10.5 eV) compares favorably with the bulk bandwidth (~11.3 eV). Recall the point we made about slow convergence with the cluster size! SRM then calculated the oxygen-related structure for several positions of oxygen atom on Al(111) surface.

At Z_0=0.53 Å SRM found the peaks at -10.0 and -7.6 eV for the A site, while for the B site the peaks are at -9.6 and -7.8 eV. Both of these peaks have their origin in the $p_{x,y}$ orbitals parallel to the surface. The features due to the p_z orbitals were found at lower binding energies giving rise to peaks at -5.4, -3.5, and -0.5 eV for the A site and at -5.9, -3.5, and -0.5 eV for the B site. SRM argued that the actual distance of oxygen on Al(111) may be somewhat larger because that would remove the peak around -5.5 eV to lower binding energy as it has not been seen experimentally. So they concluded that Z_0 ought to be somewhere between 0.53 and 1.06 Å. The oxygen-related structure then should be found around -10 and -7 eV, and between -5 and E_F. The current experimental situation is that the two higher binding energy peaks have been observed in ARUPS [3.97-99], but no oxygen-related structure has been seen between -5 and 0. A surface EXAFS experiment [3.84] has given a value of $0.55<Z_0<0.80$ Å in good agreement with the SRM cluster results. However, a recent LEED measurement [3.83] has given a value of Z_0=1.33 Å, which is in substantial disagreement with the above findings. Earlier, LANG and WILLIAMS [3.48] obtained Z_0=1.32 Å [3.49] when the discrete aluminum lattice was included. A recent ellipsometric study [3.86] indicates a value 2 Å and it is fair to say that the problem has not been fully solved yet.

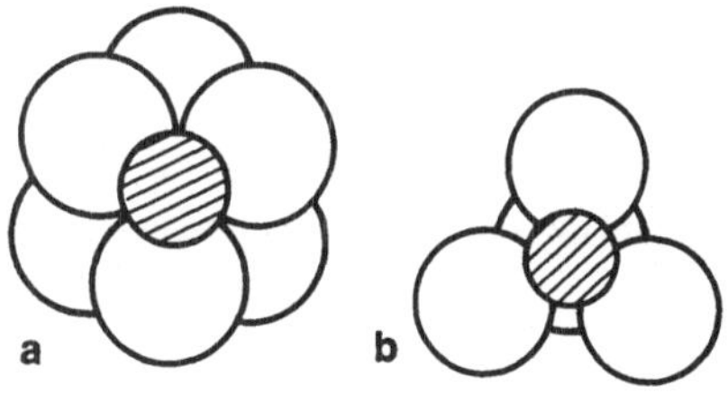

Fig. 3.5a,b. Two distinct threefold chemisorption sites for O on Al(111). In (a) the adsorbate is in a threefold coordinated site with no second-layer Aℓ atom directly below (an FCC site), while in (b) there is an Aℓ atom directly below (an HCP site)

3.3.2 Band-Structure Approach and Experimental Results

If adsorbate-adsorbate interactions are important, as is the case for ordered overlayers, the electronic structure is best described in terms of electronic bands. Ordered overlayers have two-dimensional periodicity and one can calculate the dependence of energy on the wavevector $\underline{k}_{\parallel}$ of the two-dimensional surface Brillouin zone (SBZ). The band dispersion (E vs $\underline{k}_{\parallel}$) can be related to peak positions in an angular-resolved photoemission experiment [3.97-99]. As alluded to earlier, it is not conclusively established that oxygen forms a stable overlayer on all of the low-index crystal surfaces. There is some evidence which suggests [3.79,82-83] that on the Al(111) face an ordered overlayer of oxygen atoms exhibiting (1×1) symmetry is developed at monolayer coverage.

Theoretical calculations have been performed for ordered monolayers on Al(001) as well as on Al(111) surfaces with qualitatively similar results. We now discuss in some detail the results of band-structure calculations for the oxygen-aluminum system. As the basis of our ideas we follow the work of BATRA and CIRACI (BC) [3.56], which is a slab calculation using the ETB-xα method modeling an ordered (1×1) oxygen monolayer on Al(001). To systematically understand the nature of interaction between the substrate and overlayer, these authors first calculated the two-dimensional band structure for an isolated layer of oxygen atoms in a square lattice of side 2.86 Å appropriate for Al(001). The dispersion of the three oxygen p bands along various directions in the SBZ is shown in Fig.3.6a. At Γ, the p_z band (taking z as the direction of the surface normal) lies below the doubly degenerate p_x,p_y states because the latter form an antibonding combination at $\underline{k}_{\parallel}=0$. As one moves from Γ towards X, the p_x band becomes more and more bonding and its energy is gradually lowered. The p_y band is antibonding here and has the highest energy. The p_z band is nonbonding in practically all directions and is relatively dispersionless. Along the XK direction, the p_y band becomes bonding and starts to dip in energy, but p_x becomes less bonding. At K, the two bands become degenerate and continue to be nearly degenerate along the KΓ direction because $p_x \pm p_y$ combinations have about equal energies. The total bandwidth is also a function of the exchange parameter α and increases to 2.2 eV for α=0.7. However, our main conclusions are independent of the choice of α. Band dispersions reported above for the oxygen overlayer are in good qualitative agreement with those calculated by other workers [3.58-59,101-102].

Next BC performed the band calculation for the overlayer in the presence of the substrate potential. The results (using an arbitrary energy scale) are illustra-

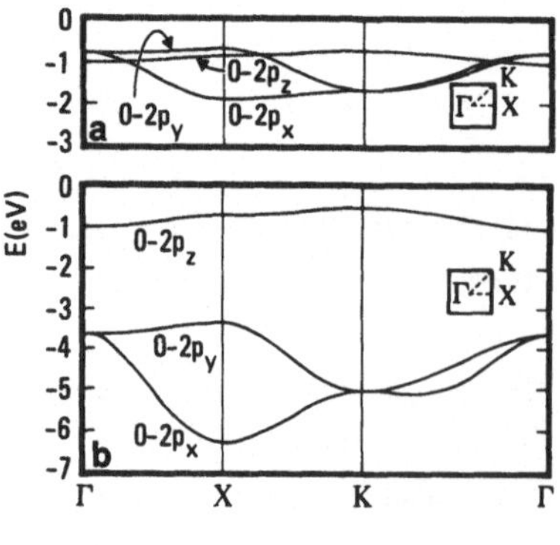

Fig. 3.6a,b. Energy bands for a two-dimensional monolayer of oxygen (2p region) (a) in the absence and (b) in the presence of the crystal potential of the eight layers of aluminum. The inset shows the SBZ (reproduced from Ref. [3.56]).

ted in Fig.3.6b. By comparing Figs.3.6a,b one notices that the total width of the oxygen p_x,p_y bands has increased by about 2 eV. The nondispersive p_z orbitals are located at higher energy. Thus the interaction with the substrate increases the binding energy and dispersion of the $p_{x,y}$ orbitals.

Figures 3.7a-c show the band structure for an eight layer slab of aluminum with chemisorbed oxygen. Oxygen monolayer bands in the crystal potential alone (Fig.3.6b) are superimposed (dotted lines). The local and orbital density of states are shown in Fig.3.7d. The peak at about -6.5 eV and the shoulder at about -8.2 eV are due to $O(2p_{x,y})$ orbitals. The $O(2p_z)$ orbitals, though dispersed in the energy range extending from E_F to about -6 eV, give an identifiable peak at about -5 eV. From these calculations one thus expects three oxygen-2p-related structures. The one with the lowest binding energy is due to $O(2p_z)$ orbitals and the other two (with higher binding energy) have their origin in $O(2p_{x,y})$ orbitals. These three peaks have been observed in angular-resolved photoemission experiments. As noted during the discussion of the cluster results, when complete adlayer interactions are considered, $O(2p_z)$ character does indeed appear at lower binding energy than the $O(2p_{x,y})$ orbitals. It is evident from the position of E_F in Fig.3.7d that in addition to $O(2p_{x,y})$, a large fraction of the $O(2p_z)$ orbitals are also occupied. The

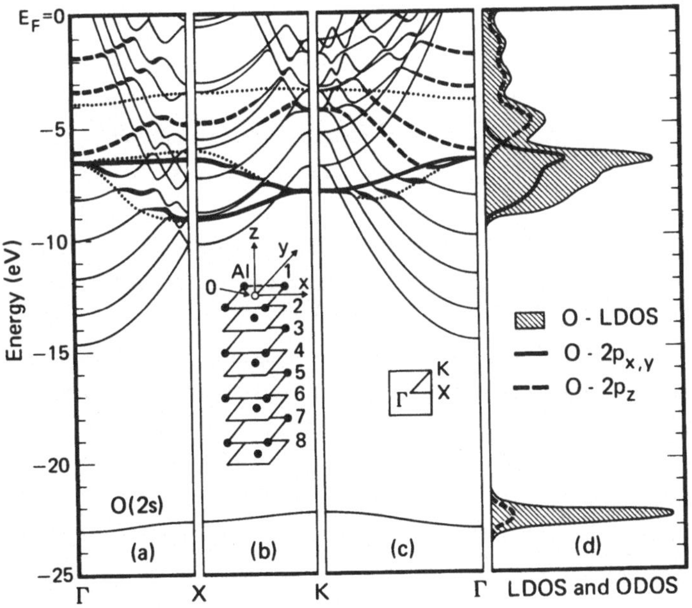

Fig. 3.7a-d. Energy bands for a monolayer of oxygen and eight layers of aluminum along Γ-X-K-Γ. Dotted lines give the bands for oxygen monolayer in the crystal field. The insets shows the chemisorption model and the BZ, (d) local and orbital densities of states for oxygen (from [3.56])

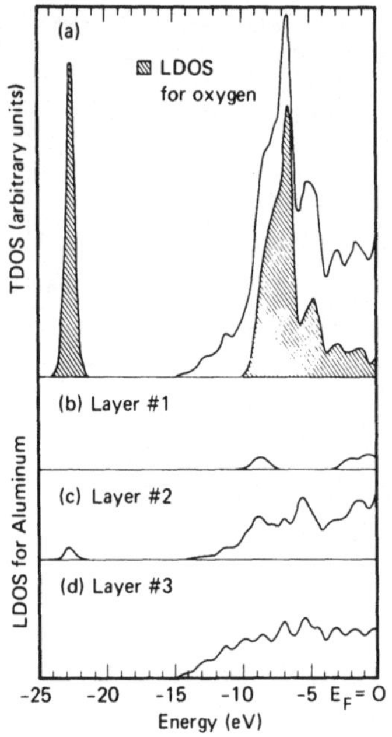

Fig. 3.8. TDOS for oxygen and three layers of aluminum for O-Al(001). Shaded area in (a) gives LDOS for oxygen. Remaining panels gives LDOS for various layers of aluminum

system may thus be thought of as "ionically" bound. A gross MULLIKEN-type population analysis [3.103] revealed that approximately 1.5 electrons have been transferred from the substrate to oxygen.

In Fig.3.8 we have plotted the TDOS and LDOS for various layers of aluminum. Figure 3.8a includes TDOS of oxygen and three layers of aluminum atoms. LDOS for oxygen is given by the shaded area. It reproduces reasonably well the broad O(2p) resonance observed at about -7 eV below E_F in photoemission experiments [3.49,72-79]. The width of this peak arises primarily from interaction of oxygen with oxygen atoms in other unit cells, namely, the band-structure effects. Thus for the purposes of studying interaction of oxygen with the substrate it suffices to focus on a few layers of Al.

LDOS corresponding to layer #1 of Al (Fig.3.8b) shows a considerable depletion of density of states. Since this layer lies in the same plane as oxygen, we conclude that charge is transferred from aluminum to oxygen. The LDOS of layer #2 also shows signs of interaction with oxygen. But layer #3 almost looks like a "bulk layer" with somewhat broadened valence band.

PAINTER [3.58] has presented a detailed theoretical study of the bonding of oxygen to Al(001) and has determined the relative roles of local and delocalized electron effects. He concluded, in agreement with other cluster calculations, that the position of main oxygen resonance peak (~7 eV below E_F) can be calculated from a "surface molecule" cluster model because it largely involves local bond mechanism of $O(2p_{x,y})$ orbitals with the substrate orbitals. The band structure for oxygen monolayer as calculated by PAINTER places the $O(2p_z)$ bands at somewhat higher binding energy when compared with LIEBSCH [3.101] or BATRA and CIRACI [3.56]. A similar trend continues when the substrate potential and hybridization effects are included. Apart from this minor difference, the results obtained by PAINTER agree quite well with the findings of BATRA and CIRACI discussed above.

BULLETT [3.59] has calculated the band structure for a monolayer of oxygen chemisorbed on Al(111) surface using a seven-layer-thick slab. Oxygen atoms were placed on every threefold hollow site of the FCC type (no aluminum atom in the secon layer below oxygen). Oxygen atoms were located outside the top layer such that d(Al-O)=1.9 Å. BULLETT showed that the width of oxygen 2p resonance increased by about an electron volt upon incorporating O-Aℓ interactions. He found isolated oxygen monolayer width to be 1.9 eV which became 3 eV when O-Aℓ interactions were included. This width is in good agreement with the experimentally observed resonance, but its location with respect to E_F is off by several electron volts.

Dispersion and symmetry of O(2p) derived bands on Al(111) surface has been investigated [3.97-99] using polarized light and angle-resolved photoemission spectroscopy (ARUPS). The first angular-resolved photoemission study of an ordered (1×1) overlayer by MARTINSON and FLODSTRÖM [3.97] used unpolarized light incident close to the surface normal. Band positions were recorded as a function of polar angle of emission, θ_e, from which E versus $k_\parallel$ plots were generated using the relation

$$k_\parallel(\text{Å}^{-1}) = 0.531[E_{kin}(eV)]^{\frac{1}{2}} \sin\theta_e \quad . \tag{3.53}$$

Various directions in the SBZ were selected (i.e., the direction of $\underline{k}_\parallel$) by changing the azimuthal angle ϕ. The experimental results of MARTINSON and FLODSTRÖM are reproduced in Fig.3.9. It shows a single *symmetric* peak at θ_e=0 (normal emission) whose halfwidth is minimum at this angle of emission. These observations were interpreted by noting that only the p_x and p_y oxygen orbitals contribute to the O(2p) surface resonance at normal emission, i.e., at the Γ point in the SBZ, while transitions involving the p_z orbitals are forbidden by the dipole selection rules. The halfwidth of the photoemitted electrons attains its minimum value for normal emission because the p_x and p_y levels are degenerate at Γ.

It is also clear from Fig.3.9 that with increasing polar angles θ_e the O(2p) peak broadens, develops more structure, and shows different degrees of splitting depending on the azimuthal angle ϕ. Most pronounced splitting (~1.5 eV) has been observed close to the $\bar{K}$ point. The suggested symmetry designations for the two peaks were p_z and (p_x,p_y), respectively, though individual peaks were not assigned. Along the $\overline{\Gamma M}$ direction, no explicit splitting has been observed, but the peak shifted to higher binding energy with increasing θ_e. This is consistent with the theoretical studies if we recall that p_x and p_y form an antibonding combination at $\bar{\Gamma}$ which becomes more bonding as one moves towards the zone edge.

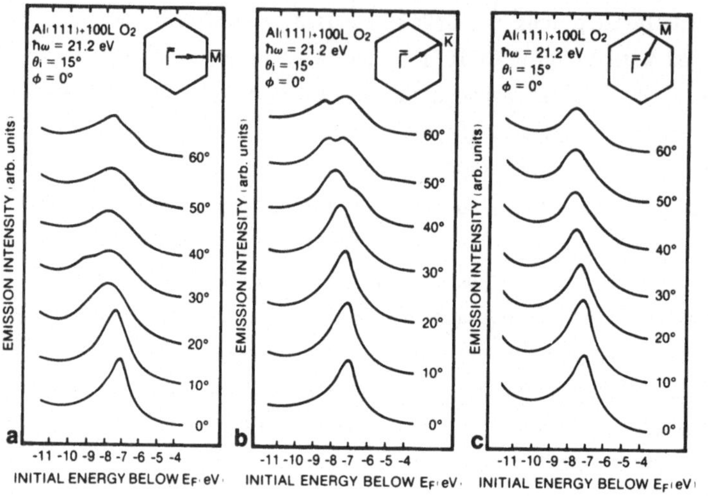

Fig. 3.9a-c. Polar-angle dependence of the O on Al(111) at $\theta_i=15^o$ and $\hbar\omega=21.2$ eV along the azimuthal directions (a) [$2\bar{1}\bar{1}$], i.e., $\phi=0^o$, (b) [$10\bar{1}$], i.e., $\phi=30^o$, and (c) [$11\bar{2}$], i.e., $\phi=60^o$ (from [3.97])

The experimental photoemission spectra is found to be different along the direction $\overline{\Gamma M}$, [$2\bar{1}\bar{1}$] and $\overline{\Gamma M}$ [$11\bar{2}$]. This has been explained in terms of reduction of the sixfold symmetry of the isolated layer to the threefold symmetry due to the substrate. Based on this azimuthal dependence the authors [3.97] suggested that the oxygen atoms are chemisorbed in the threefold symmetric FCC sites on Al(111).

An independent ARUPS study of O-Al(111) which appeared concurrently is due to EBERHARDT and HIMPSEL (EH) [3.98]. Figure 3.10 shows a typical set of energy distribution curves which they obtained at an emission angle of 30° away from the normal in the ($1\bar{1}0$) mirror plane that corresponds to a $k_{\parallel}$ near the zone boundary (M-point) of the SBZ. A single peak (dashed line) at about -8 eV below E_F was detected when the electric field was polarized perpendicular to the mirror plane of detection. On the other hand, two peaks located at -6.5 and -9.4 eV (solid line) are seen for the polarization vector in the mirror plane. This polarization dependence is expected for an ordered overlayer because the O($2p_z$) level form one even band and the O($2p_x$, $2p_y$) form two bands, one odd and the other even with respect to the mirror plane. Since the matrix elements is invariant under reflection about the mirror plane, odd initial states are seen by the electric field vector $\underline{E}$ perpendicular to the ($1\bar{1}0$) plane, and even states by $\underline{E}$ in the ($1\bar{1}0$) plane.

Complete band dispersion, shown in Fig.3.11, was obtained by changing the collection angle. In agreement with the BC calculation [3.56], the p_z state has the lowest binding energy and smallest dispersion. The highest binding energy valence band has been identifeied as the $p_{x,y}$ band which has even symmetry relative to the Γ-M axis. The band with the intermediate binding energy is then clearly the $p_{x,y}$

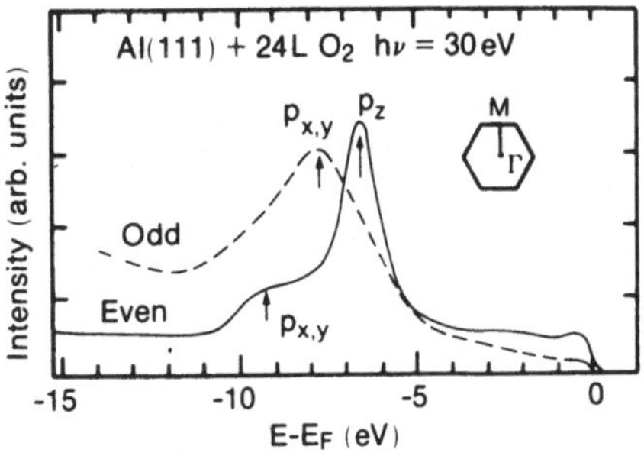

Fig. 3.10. ARUPS data showing the odd (dashed) and even (solid) oxygen induced states. The inset shows the SBZ (from [3.98])

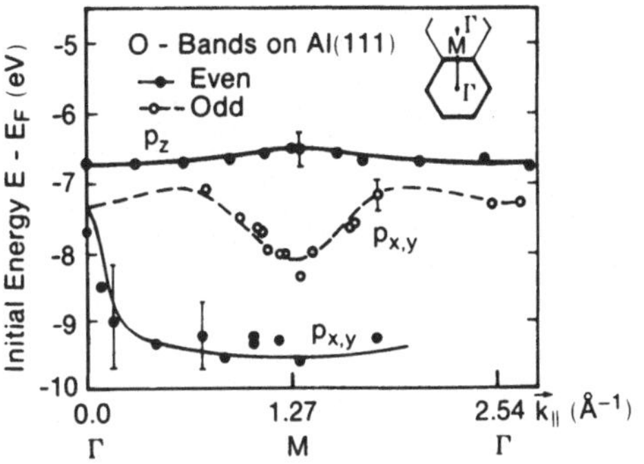

Fig. 3.11. Measured two-dimensional band dispersion of O-Al(111) (from [3.98])

odd band. There is some uncertainty in locating the three bands at Γ, because experimentally a single broad asymmetric peak of 2.5 eV FWHM is seen. EH resolved this peak by curve fitting into two Lorentzians centered at -6.7 and -7.7 eV. The -6.7 eV Lorentzian was interpreted to be the emission form p_z at Γ. The -7.7 eV Lorentzian was assigned to the emission from the two $p_{x,y}$ bands which by theoretical considerations have to be degenerate at Γ. This is to be compared with the independently measured value (-7.3±0.3 eV) for the Γ point of the second zone.

It is reasonable to state that the present experiment has verified, at least qualitatively, most of the features calculated for the Al(001) surface. The calculation showed that the lateral interaction between oxygen atoms would result in a higher binding energy for the p_z level than for the $p_{x,y}$ levels at Γ. The BC calculation further showed that the substrate interaction would increase the binding energy and dispersion of the $p_{x,y}$ orbitals. This is precisely what EH have observed for the Al(111) surface. The degenerate $p_{x,y}$ orbitals have a higher binding energy at Γ than the p_z orbital. The $p_{x,y}$ bands have a dispersion ~2.7 eV. This dispersion is mainly determined by oxygen-substrate-oxygen interaction via the $p_{x,y}$ orbitals. From this, EH concluded that oxygen initially chemisorbs in an ordered fashion on the threefold hollow FCC position on the Al(111) surface.

HOFMANN et al. [3.99] have also recently observed dispersion effects on the oxygen 2p-derived levels using ARUPS. They have identified the symmetries of various bands using polarized HeI radiation. An important new ingredient of their work consisted of showing that the ordered overlayers grew via an island growth mechanism and that it transformed irreversibly into disordered structure within 1 - 2 h. Such an instability of the (1×1) layer is clearly illustrated in Fig.3.12 which shows spectra recorded immediately after oxygen exposure (A) and one 60 min. later (B). A general loss of clear features in going from A to B indicates a transformation into a less well-ordered structure. The above authors also performed experiments at various oxygen exposures. They noted that the three peaks shown in A were

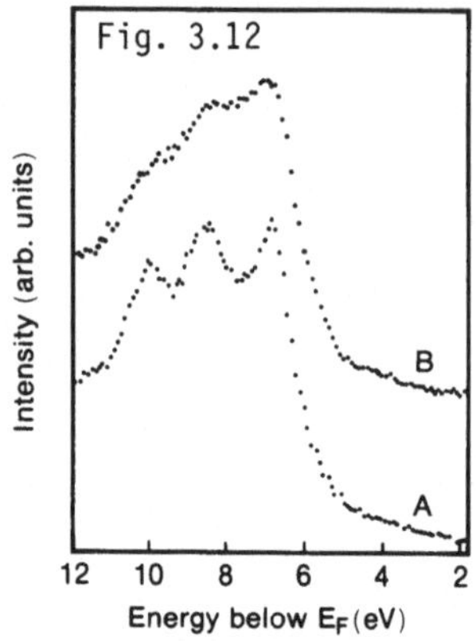

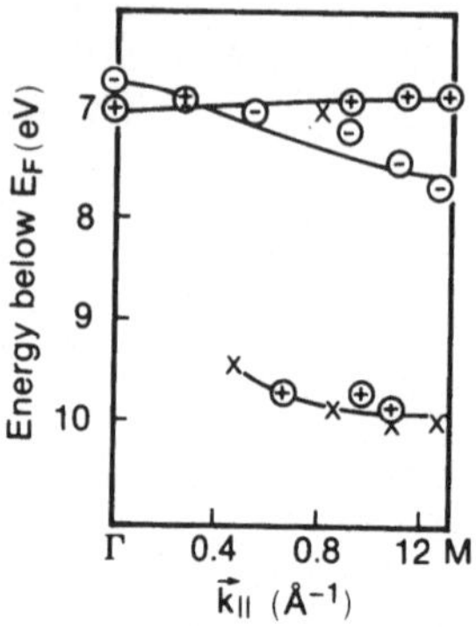

Fig.3.12. Effect of elapsed time after exposure on the appearance of O-Aℓ(111) spectra. Curve A, spectra recorded immediately after oxygen exposure. Curve B, 60 min. later (from [3.99]).

Fig.3.13. Experimental band structure for O-Aℓ(111) along the ΓM direction. See text for explanation (from [3.99])

present with constant separation from the very lowest converges. From this they concluded that the (1×1) layer is formed via an island growth mechanism.

The experimental band structure obtained by HOFMANN et al. along the ΓM direction is reproduced in Fig.3.13. The symbols ⊕, ⊖ and x indicate peaks in spectra recorded with p-polarized, s-polarized, and unpolarized light, respectively. The letters s and p refer to the $\underline{E}$ vector oriented normal (s) or parallel (p) to the plane of incidence. States antisymmetric with respect to the mirror plane contribute to spectra with s-light, while p-light excites electrons from symmetric states. The overall dispersion and symmetry assignements of various bands agree with the result reported by EH. At the Γ point, however, the two groups have different symmetry assignments. HOFMANN et al. [3.99] place the symmetric p_z-derived band at higher binding energy than the antisymmetric $p_{x,y}$-derived band at Γ. They rationalize this interpretation by noting that in the layer, the p_z band is completely bonding at Γ, while the $p_{x,y}$ band is mainly antibonding. EH [3.98] label the bands in the reverse order and point to the calculation by BC [3.56] for Al(001). The calculation shows that p_z band, which is antibonding in the layer and has higher binding energy, does in fact move to lower binding energy when substrate effects are included. Thus the ordering of the bands at Γ is not resolved yet because a single broad peak is observed near normal emission, whose decomposition into components bands is problematical.

3.3.3 Unresolved Issues

Having described the knowledge that has been acquired about the oxygen-aluminum system, let us briefly mention the issues which have not been resolved satisfactorily thus far.

Table 3.1. A summary of reported values for the oxygen-aluminum interplanar distance (Z_0) and oxygen-aluminum bond length (d) for O/Al(111)

Technique and Source	Z_0[Å]	d[Å]
Jellium Theory [3.48]	1.75	2.40
Jellium with lattice corrections [3.49]	1.32	2.11
SCF-Xα Molecular Cluster Theory [3.52]	0.53-1.06	1.73-1.96
LEED [3.83]	1.33±0.08	2.12±0.05
SEXAFS [3.84]	$0.70^{+0.10}_{-0.15}$	$1.79^{+0.04}_{-0.05}$
LEED [3.117]	1.46±0.05	2.21±0.03
SCF-LAPW Band-Structure Theory [3.118]	~0.70	~1.79
HF-Molecular Cluster Theory [3.119]	0.70±0.10	1.79±0.05
Photoemission and SEXAFS [3.120]	0.98±0.10	1.92±0.05
SCF-LCAO Band-Structure Theory [3.121]	0.55	1.74

1) How does the oncoming oxygen molecule break up into atomic species which subsequently chemisorp or diffuse into the bulk?

2) Is the ordered oxygen overlayer formed only on Al(111) and why? What are the geometrical parameters? (See Table 3.1)

3) Why there is no oxygen-induced emission in the 2 - 5 eV range below E_F on the Al(111) surface? Calculationally such an emission has been predicted for Al(001) as well as for Al(111) but apparently it has only been seen for Al(001).

4) Is the peak 9 - 10 eV below E_F truly related to the existence of an ordered (1×1) chemisorbed layer, or is the alternative explanation, namely, the co-existence of chemisorbed and incorporated phases, more appropriate?

5) What is the proper ordering of oxygen bands at the Γ point?

6) Does the work function increase or decrease upon oxygen chemisorption?

7) What is the origin of the 1.4 eV chemical shift of the Aℓ 2p core level upon exposure to oxygen?

Thus it seems that the oxygen-aluminum system shall remain an active area of research for the next several years. Hopefully, the detailed information gained on this system shall be readily transferrable to transition metal surfaces which are of more practical interest.

Acknowledgements. The work on embedded cluster was completed at Orsay, France during the CECAM workshop on "Computational Methods for Small Molecules Adsorbed on Solid Surfaces." The author is grateful to C. Pisani, A. van der Avoird, J.P. Muscat and

J.L. Whitten for stimulating discussions on the subject. The author would also like to thank Dr. C. Moser for an invitation to the CECAM workshop and to IBM Europe for their kind hospitality. The author appreciates the careful reading given to this chapter by Dr. W.H. Marlow.

Appendix 3.A Electron Spectroscopies for Surface Analysis [3.104]

Name	Acronym	Incident Particle	Incidence Energy	Emitted Particle	Emitted Energy
X-ray Photoemission Spectroscopy	XPS (ESCA)	Photon (X-ray)	High (1-10 keV)	Electron	0-10 keV
Ultra Violet Photo-emission Spectroscopy	UPS	Photon (UV)	Low (4-40 eV)	Electron	0-30 eV
Low-Energy Electron Diffraction (Elastic)	LEED (ELEED)	Electron	15-500 eV	Incident Electron	Within 0.5 eV of primary energy
Inelastic Low-Energy Electron Diffraction	ILEED	Electron	1-200 eV	Incident Electron	Losses 5-50 eV
Auger Electron Spectroscopy	AES	Electron	100-5000 eV	Electron	20-2000 eV
Scanning Auger Electron Spectroscopy	SAES	Electron	3-15 keV	Electron	10-1000 eV
Auger Electron Appearance Potential Spectroscopy	AEAPS	Electron	100-2000 eV	Electron	50-1400 eV
Ionization (Loss) Spectroscopy	IS (ILS)	Electron	250-1000 eV	Incident Electron	100-500 eV
Low-Energy Energy Loss Spectroscopy (Characteristic)	LEELS (CELS)	Electron	Low (1-50 eV)	Incident Electron	Losses 0.02-a few eV
Transmission Energy Loss Spectroscopy (Characteristic)	TELS (CELS)	Electron	High (20-60 keV)	Incident Electron	Losses 0.005-a few eV
Field Emission Spectroscopy	FES	----	----	Electron	~2 keV
Ion Neutralization Spectroscopy	INS	Ion (He^+, Ne^+)	4-10 eV	Electron	0-16 eV
Electron Stimulated Desorption Ion Angular Distribution	ESDIAD	Electron	50-2000 eV	Ion	0-10 eV
Photon Stimulated Desorption Ion Angular Distribution	PSDIAD	Photon	15-100 eV	Ion	0-10 eV
Surface Extended X-ray Absorption Fine Structure	SEXAFS	Photon	300-1000 eV	Electron	0-1000 eV
Extended Appearance Potential Fine Structure	EAPFS	Electron	500-1500 eV	Electron	0-500 eV

Appendix 3.B Evaluation of Matrix Elements of Various Operators for Gaussian Basis

A general Gaussian function (α) centered at site $\underline{A}$ is defined as

$$(\alpha) \equiv (A,\alpha,\ell,m,n) \equiv N(x-A_x)^\ell (y-A_y)^m (z-A_z)^n \exp(-\alpha r_A^2)$$

$$= N x_A^\ell y_A^m z_A^n \exp(-\alpha r_A^2) \quad . \tag{3.54}$$

The *normalization constant* N is given by

$$N = \left[\left(\frac{\pi}{2}\right)^{1.5}\left(\frac{1}{\alpha}\right)^{1.5} \frac{(2\ell-1)!!(2m-1)!!(2n-1)!!}{2^{2(\ell+m+n)}\alpha^{(\ell+m+n)}}\right]^{-\frac{1}{2}} \quad , \tag{3.55}$$

where the double factorial, $(2k-1)!! = 1$ if $k = 0$ and $(2k-1)!! = (2k-1)!/[2^{k-1}(k-1)!]$ if $k > 0$.

The Overlap integral between Gaussian functions $(\alpha)_1$ and $(\alpha)_2$ is

$$\langle(\alpha)_1|(\alpha)_2\rangle \equiv \langle A,\alpha_1,\ell_1,m_1,n_1|B,\alpha_2,\ell_2,m_2,n_2\rangle$$

$$= N_1 N_2 \left(\frac{\pi}{\alpha_1+\alpha_2}\right)^{1.5} \exp\left(-\frac{\alpha_1\alpha_2}{\alpha_1+\alpha_2}\overline{AB}^2\right) S_x S_y S_z \quad , \tag{3.56}$$

where

$$S_x = \sum_{k=0}^{[1/2(\ell_1+\ell_2)]} f_{2k}(\ell_1,\ell_2,\overline{DA}_x,\overline{DB}_x) \frac{(2k-1)!!}{2^k(\alpha_1+\alpha_2)^k} \quad , \tag{3.57}$$

with similar definitions for S_y and S_z. Here [m] denotes largest integer $\le$ m. Note that $S_x = 1$ if $\ell_1 = \ell_2 = 0$. Furthermore, $f_j(\ell,m,a,b)$ means the coefficient of x^j in the expansion of x^j in the expansion of $(x+a)^\ell (x+b)^m$, and

$$\underline{D} = (\alpha_1\underline{A}+\alpha_2\underline{B})/(\alpha_1+\alpha_2) \quad , \tag{3.58}$$

$$\overline{AB}^2 = (A_x-B_x)^2 + (A_y-B_y)^2 + (A_z-B_z)^2 \quad , \tag{3.59}$$

$$\overline{DA}_x = D_x - A_x = \frac{\alpha_1 A_x + \alpha_2 B_x}{\alpha_1+\alpha_2} - A_x \quad , \tag{3.60}$$

$$\overline{DB}_x = D_x - B_x = \frac{\alpha_1 A_x + \alpha_2 B_x}{\alpha_1+\alpha_2} - B_x \quad , \tag{3.61}$$

and $\overline{DA}_y$, $\overline{DA}_z$, $\overline{DB}_y$, $\overline{DB}_z$ are similarly defined.

The kinetic energy integral between two Gaussian functions can be expressed in terms of overlap integrals as stated below:

$$N^{i}_{N,N_2} <(\alpha)_1|-\nabla^2|(\alpha)_2> = 2\alpha_2[2(\ell_2+m_2+n_2)+3] <(\alpha)_1|(\alpha)_2> - 4\alpha_2^2\{<(\alpha)_1|B,\alpha_2,\ell_2+2,m_2,n_2>$$

$$+ <(\alpha)_1|B,\alpha_2,\ell_2,m_2+2,n_2> + <(\alpha)_1|B,\alpha_2,\ell_2,m_2,n_2+2>\}$$

$$- \{\ell_2(\ell_2-1)<(\alpha)_1|B,\alpha_2,\ell_2-2,m_2,n_2>$$

$$+ m_2(m_2-1)<(\alpha)_1|B,\alpha_2,\ell_2,m_2-2,n_2>$$

$$+ n_2(n_2-1)<(\alpha)_1|B,\alpha_2,\ell_2,m_2,n_2-2>\} \quad . \tag{3.62}$$

In the evaluation of the *core part of the crystal potential*, we encounter matrix elements of the type $<(\alpha)_1|\exp(-\beta r_C^2)|(\alpha)_2>$ and $<(\alpha)_1[\exp(-\beta r_C^2)]/r_C|(\alpha)_2>$, where C denotes the coordinates of various lattice sites of the system. These can be easily evaluated as follows:

$$<(\alpha)_1|\exp(-\beta r_C^2)|(\alpha)_2> = N_1N_2\left(\frac{\pi}{\alpha_1+\alpha_2+\beta}\right)^{1.5} \exp\left[-\frac{\alpha_1\alpha_2}{\alpha_1+\alpha_2}\overline{AB}^2 - \frac{(\alpha_1+\alpha_2)\beta}{\alpha_1+\alpha_2+\beta}\overline{DC}^2\right]$$

$$\times T_xT_yT_z \quad , \tag{3.63}$$

where

$$T_x = \sum_{k=0}^{[1/2(\ell_1+\ell_2)]} f_{2k}(\ell_1,\ell_2,\overline{PA}_x,\overline{PB}_x) \frac{(2k-1)!!}{2^k(\alpha_1+\alpha_2+\beta)^k} \quad , \tag{3.64}$$

$$\underline{P} = \frac{\alpha_1\underline{A}+\alpha_2\underline{B}+\beta\underline{C}}{\alpha_1+\alpha_2+\beta} \quad , \tag{3.65}$$

$$\overline{DC}^2 = (D_x-C_x)^2 + (D_y-C_y)^2 + (D_z-C_z)^2 \quad , \tag{3.66}$$

and where $\underline{D}$ is defined in (3.58), $\overline{PA}_x = P_x - A_x$, etc. The matrix elements for the *damped nuclear attraction term* are given by

$$<(\alpha)_1|\frac{\exp(-\beta r_C^2)}{r_C}|(\alpha)_2> = N_1N_2\,\frac{2\pi}{\gamma} \times \exp\left[-\frac{\alpha_1\alpha_2}{\alpha_1+\alpha_2}\overline{AB}^2 - \frac{(\alpha_1+\alpha_2)\beta}{\gamma}\overline{DC}^2\right]$$

$$\times \sum_{i,r,u} A_{i,r,u}(\ell_1,\ell_2,A_x,B_x,C_x,\gamma)$$

$$\times \sum_{j,s,v} A_{j,s,v}(m_1,m_2,A_y,B_y,C_y,\gamma)$$

$$\times \sum_{k,t,w} A_{k,t,w}(n_1,n_2,A_z,B_z,C_z,\gamma)F_\nu(\gamma\overline{PC}^2) \quad , \tag{3.67}$$

where $\gamma = \alpha_1 + \alpha_2 + \beta$, $\nu = i + j + k - 2(r+s+t) - (u+v+w)$, and

$$A_{i,r,u} = (-)^i f_i(\ell_1,\ell_2,\overline{PA}_x,\overline{PB}_x)\,\frac{(-)^u i!(\overline{PC}_x)^{i-2r-2u}(1/4\gamma)^{r+u}}{r!u!(i-2r-2u)!} \quad . \tag{3.68}$$

$A_{j,s,v}$, $A_{k,t,w}$ are similarly defined in terms of the y and z components and the incomplete gamma function

$$F_\nu(a) = \int_0^1 x^{2\nu}\exp(-ax^2)dx, \quad (t>0, \quad \nu=0,1,2...) \quad . \tag{3.69}$$

The summations in (3.67) are over the following ranges

$$i=0\to\ell_1+\ell_2, \quad r=0\to[i/2], \quad u=0\to[(i-2r)/2] \quad ,$$

$$j=0\to m_1+m_2, \quad s=0\to[j/2], \quad v=0\to[(j-2s)/2] \quad ,$$

$$k=0\to n_1+n_2, \quad t=0\to[k/2], \quad w=0\to[(k-2t)/2] \quad . \tag{3.70}$$

The matrix elements of the *smooth part* of the crystalline potential requires complex conjugate of the expression

$$<(\alpha)_1|e^{i\underline{G}\cdot\underline{r}}|(\alpha)_2> = N_1N_2\,\exp\left(i\underline{G}\cdot\underline{D}-\frac{G^2}{4(\alpha_1+\alpha_2)}-\frac{\alpha_1\alpha_2}{\alpha_1+\alpha_2}\overline{AB}^2\right)\times Q_xQ_yQ_z \quad , \tag{3.71}$$

where

$$Q_x = \sum_{k=0}^{\ell_1+\ell_2} f_k(\ell_1,\ell_2,\overline{DA}_x,\overline{DB}_x)\left(\frac{1}{2\sqrt{\alpha_1+\alpha_2}}\right)^k H_k\left(\frac{G_x}{2\sqrt{\alpha_1+\alpha_2}}\right)i^k \quad , \tag{3.72}$$

with similar expressions for Q_y, Q_z in terms of y and z components. The function H_k is given by

$$H_k(Z) = k!\sum_{j=0}^{[k/2]}\frac{(-)^j(2Z)^{k-2j}}{j!(k-2j)!} \quad , \tag{3.73}$$

$H_0(Z)=1$, and $H_k(o)=0$ if k is an odd integer, $H_k(o)=[(-)^{k/2}(k)!]/(k/2)!$ if k is an even integer. In (3.71,72), $i=\sqrt{-1}$. During the evaluation of these expressions, one stores complex and imaginary parts for eventual extraction of the real part of the matrix elements. One can readily verify that in the limit $\underline{G}\to 0$, (3.71) reduces to the expression for the overlap integral as given in (3.56).

Appendix 3. C On Fourier Transforming the Local Exchange Potential

One of the crucial steps in numerically computing the electronic band structure is the ability to Fourier transform (FT) the local exchange potential. Since the local exchange potential is a nonlinear function of the total charge density, none of the analytical techniques are applicable. Various authors have resorted to special methods which work particularly well for simpler cases, e.g., one atom per unit cell. The fast Fourier transform (FFT) techniques work well if the charge density is extremely smooth and slowly varying. Furthermore, to utilize the full power of FFT, one must choose 2^n points along each dimension. This is often inconvenient and quickly leads to a huge mesh, especially when one realizes that we are to Fourier transform in three dimensions. Furthermore, the exponent n also limits the maximum number of Fourier coefficients one can request. Thus we feel that a modification is still needed. We discuss below a scheme which is general and has worked well for us in calculating the Fourier transform of the exchange potential for use in electron structure computation of solids and surfaces. In the interest of generality we have not assumed any special symmetry properties for the atoms in the unit cell. It is obvious that some symmetry relations can be easily implemented to save factors of two in actual computation. The FT of the exchange potential is given by

$$V_x(\underline{G}) = \frac{1}{\Omega}\int_{\Omega} V_x(\underline{r})\,\exp(i\underline{G}\cdot\underline{r})d\underline{r} \quad , \tag{3.74}$$

where the integration is confined to the reference unit cell of volume Ω and $\underline{G}$ is a reciprocal lattice vector. In the overlapping charge density approximation, the local exchange potential is

$$V_x(\underline{r}) = -3\alpha\left[\frac{3}{\pi}\rho_T(\underline{r})\right]^{1/3} = C\rho_T^{1/3}(\underline{r}) \quad , \tag{3.75}$$

where α is the exchange parameter, $C = -3\alpha(3/\pi)^{1/3}$ and

$$\rho_T(\underline{r}) = \sum_{\mu,\nu} \rho_\mu(\underline{r}-\underline{R}_\nu-\underline{\tau}_\mu) \tag{3.76}$$

is the total charge density (taken as the sum of ρ_μ's, the atomic charge densities). Further, $\underline{R}_\nu$ are the direct-lattice translation vectors, and $\underline{\tau}_\mu$ are nonprimitive translation vectors for the atoms in the unit cell.

To proceed with the integration in (3.74), we have to overcome two major obstacles.

1) The exchange potential near each nuclear site is quite large and drops quite rapidly as one moves away from the core region. Thus the integrand is rather ill behaved.

2) Since the integrand is oscillatory in nature, for large values of $|\underline{G}|$ an extremely fine grid is needed unless one employs some special techniques.

To handle the first problem we write (3.74) by adding and subtracting a suitably chosen set of analytical functions centered on every atom. The actual choice is described later.

$$V_x(\underline{G}) = \frac{1}{\Omega}\int_\Omega d\underline{r}\; e^{i\underline{G}\cdot\underline{r}}\Big[V_x(\underline{r}) - \sum_{\mu,\nu} f_\mu(\underline{r}-\underline{R}_\nu-\underline{\tau}_\mu)\Big] + \frac{1}{\Omega}\int_\Omega d\underline{r}\; e^{i\underline{G}\cdot\underline{r}} \sum_{\mu,\nu} f_\mu(\underline{r}-\underline{R}_\nu-\underline{\tau}_\mu)$$

$$\equiv \Delta V_x(\underline{G}) + V_x^a(\underline{G}) \quad , \tag{3.77}$$

where

$$\Delta V_x(\underline{G}) = \frac{1}{\Omega}\int_\Omega d\underline{r}\; e^{i\underline{G}\cdot\underline{r}}\, \Delta V_x(\underline{r}) \quad , \tag{3.78}$$

$$V_x^a(\underline{G}) = \frac{1}{\Omega}\int_\Omega d\underline{r}\; e^{i\underline{G}\cdot\underline{r}} \sum_{\mu,\nu} f_\mu(\underline{r}-\underline{R}_\nu-\underline{\tau}_\mu) \quad , \tag{3.79}$$

$$\Delta V_x(\underline{r}) = C\Big[\sum_{\mu,\nu} \rho_\mu(\underline{r}-\underline{R}_\nu-\underline{\tau}_\nu)\Big]^{1/3} - \sum_{\mu,\nu} f_\mu(\underline{r}-\underline{R}_\nu-\underline{\tau}_\mu) \quad . \tag{3.80}$$

At this point our hope is that we would be able to integrate (3.78) by using some numerical technique and relying on the fact that $\Delta V_x(\underline{r})$ is both smooth and numerically small. We shall discuss the implementation later. Equation (B.79) should be soluble analytically (superscript "a" stands for analytical) with an appropriate choice of the function f. However, before that can be accomplished with any function f, we must put it in a form where the integration over $\underline{r}$ spans the entire space and not just a reference cell as is implied in (3.79). Let us then rewrite (3.79) in the form

$$V_x^a(\underline{G}) = \frac{1}{\Omega}\int_\Omega d\underline{r}\; e^{i\underline{G}\cdot(\underline{r}+\underline{R}_m)} \sum_{\mu,\nu} f_\mu(\underline{r}+\underline{R}_m-\underline{R}_\nu-\underline{\tau}_\mu) \quad . \tag{3.81}$$

This is true for every value of m since the terms in $\sum f_\mu$ are the same in (3.79,81) and $\exp(i\underline{G}\cdot\underline{R}_m) = 1$ for all reciprocal lattice vectors. Therefore, if N is the number of unit cells in the crystal, we can write

$$V_x^a(\underline{G}) = \frac{1}{N\Omega}\int_\Omega d\underline{r}\; e^{i\underline{G}\cdot(\underline{r}+\underline{R}_m)} \sum_{\mu,\nu}\sum_m f_\mu(\underline{r}+\underline{R}_m-\underline{R}_\nu-\underline{\tau}_\mu) \quad .$$

But as we integrate over $\underline{r}$ in the reference unit cell and sum over all cells m, $\underline{r}+\underline{R}_m$ varies exactly once through all space. Thus

$$V_x^a(\underline{G}) = \frac{1}{N\Omega} \sum_{\mu,\nu} \int f_\mu(\underline{r}'-\underline{R}_\nu-\underline{\tau}_\mu)\, \exp(i\underline{G}\cdot\underline{r}')d\underline{r}' \quad ,$$

where the integration now extends over all space. If we redefine $\underline{r}' = \underline{r} + \underline{R}_\nu + \underline{\tau}_\mu$, we can at once show that

$$V_x^a(\underline{G}) = \frac{1}{\Omega} \sum_\mu e^{i\underline{G}\cdot\underline{\tau}_\mu} \int f_\mu(\underline{r})\, \exp(i\underline{G}\cdot\underline{r})d\underline{r} = \sum_\mu e^{i\underline{G}\cdot\underline{\tau}_\mu} f_\mu(\underline{G}) \quad , \tag{3.82}$$

where

$$f_\mu(\underline{G}) = \frac{1}{\Omega} \int f_\mu(\underline{r}) e^{i\underline{G}\cdot\underline{r}} d\underline{r} \quad . \tag{3.83}$$

Notice that if $f_\mu(\underline{r})$ is set equal to atomic density, then $f_\mu(\underline{G})$ become the usual atomic scattering (form) factors and $V_x^a(\underline{G})$ the geometrical structure factor encountered in X-ray scattering theory. If $f_\mu(\underline{r})$ is a spherically symmetric function, then the integral in (B.83) depends only on $|\underline{G}|$. In practice we choose

$$f_\mu(r) = \sum_i a_i^\mu e^{-\beta_i^\mu r^2} \quad ,$$

which gives

$$f_\mu(|\underline{G}|) = \frac{\pi^{3/2}}{\Omega} \sum_i \frac{a_i^\mu}{(\beta_i)^{3/2}} \exp(-G^2/4\beta_i^\mu) \quad .$$

Usually a_i^μ and β_i^μ are chosen to accurately represent the exchange potential within the core region (~1 Å) of each atom μ. This can be easily accomplished by some nonlinear least-square numerical techniques. It is important to note that once a_i^μ, β_i^μ are obtained for an atom, they are transferrable from one solid-state problem to another. One might even think of generating tables containing set of exponents and coefficients for all atoms in the periodic table. Thus the analytical part of the exchange potential is given by

$$V_x^a(\underline{G}) = \frac{\pi^{3/2}}{\Omega} \sum_\mu e^{i\underline{G}\cdot\underline{\tau}_\mu} \sum_i \frac{a_i^\mu}{(\beta_i)^{3/2}} e^{-G^2/4\beta_i^\mu} \quad . \tag{3.84}$$

Now let us turn our attention to (3.78) and note that

$$\underline{G}\cdot\underline{r} = 2\pi \sum_{j=1}^{3} \gamma_j \nu_j \quad ,$$

where

$$\underline{r} = \sum_{j=1}^{3} \gamma_j \underline{R}_j \quad , \quad 0 \le \gamma_j < 1 \quad ,$$

$$\underline{G} = \sum_{j=1}^{3} \nu_j \underline{G}_j \quad , \quad \nu_j \text{ are integers} \quad .$$

Here $\underline{R}_i$ are direct and $\underline{G}_i$ are primitive reciprocal lattice vectors such that $\underline{R}_i \cdot \underline{G}_j = 2\pi\delta_{ij}$. If we transformed the integration from $\underline{r}$ space to γ space, we can easily show that the Jacobian of transformation is simply Ω, the unit cell volume. Hence

$$\Delta V_x(\underline{G}) = \int_0^1 d\gamma_3 \int_0^1 d\gamma_2 \int_0^1 d\gamma_1 \, \Delta V_x(\underline{r}) \exp\left(2\pi i \sum_{j=1}^{3} \nu_j \gamma_j\right) \equiv \Delta V_x^R(\underline{G}) + i\Delta V_x^I(\underline{G}) \quad ,$$

where

$$\Delta V_x^R(\underline{G}) = \int_0^1 \int_0^1 \int_0^1 d\gamma_3 d\gamma_2 d\gamma_1 \, \Delta V_x(\underline{r}) \cos\left(2\pi \sum_{j=1}^{3} \gamma_j \nu_j\right) \quad ,$$

$$\Delta V_x^I(\underline{G}) = \int_0^1 \int_0^1 \int_0^1 d\gamma_3 d\gamma_2 d\gamma_1 \, \Delta V_x(\underline{r}) \sin\left(2\pi \sum_{j=1}^{3} \gamma_j \nu_j\right) \quad .$$

Real and imaginary parts can be handled in a similar fashion. So we proceed to illustrate the computation on the real part. Then

$$\begin{aligned}
\Delta V_x^R(\underline{G}) = & \int_0^1 d\gamma_3 \cos 2\pi\nu_3\gamma_3 \int_0^1 d\gamma_2 \cos 2\pi\nu_2\gamma_2 \int_0^1 d\gamma_1 \, \Delta V_x(\underline{r}) \cos 2\pi\nu_1\gamma_1 \\
& - \int_0^1 d\gamma_3 \cos 2\pi\nu_3\gamma_3 \int_0^1 d\gamma_2 \sin 2\pi\nu_2\gamma_2 \int_0^1 d\gamma_1 \, \Delta V_x(\underline{r}) \cos 2\pi\nu_1\gamma_1 \\
& - \int_0^1 d\gamma_3 \sin 2\pi\nu_3\gamma_3 \int_0^1 d\gamma_2 \cos 2\pi\nu_2\gamma_2 \int_0^1 d\gamma_1 \, \Delta V_x(\underline{r}) \sin 2\pi\nu_1\gamma_1 \\
& - \int_0^1 d\gamma_3 \sin 2\pi\nu_3\gamma_3 \int_0^1 d\gamma_2 \sin 2\pi\nu_2\gamma_2 \int_0^1 d\gamma_1 \, \Delta V_x(\underline{r}) \sin 2\pi\nu_1\gamma_1 \quad .
\end{aligned}$$

These oscillatory integrals can be completed by repeated applications of Filon's integration formula:

$$\int_{x_0}^{x_{2n}} f(x)\cos Qx\,dx \approx h[\alpha(Qh)(f_{2n}\sin Qx_{2n}-f_0\sin Qx_0)+\beta(Qh)C_{2n}+\gamma(Qh)C_{2n-1}] \quad ,$$

$$\int_{x_0}^{x_{2n}} f(x)\sin Qx\,dx \approx h[\alpha(Qh)(f_0\cos Qx_0-f_{2n}\cos Qx_{2n})+\beta(Qh)S_{2n}+\gamma(Qh)S_{2n-1}] \quad ,$$

where h is equally spaced interval $x_{i+1}-x_i=h$,

$$\alpha(t)=\frac{1}{t}+\frac{\sin 2t}{2t^2}-\frac{2\sin^2 t}{t^3} \quad ,$$

$$\beta(t)=2\left(\frac{1+\cos^2 t}{t^2}-\frac{\sin 2t}{t^3}\right) \quad ,$$

$$\gamma(t)=4\left(\frac{\sin t}{t^3}-\frac{\cos t}{t^2}\right) \quad ,$$

$$C_{2n}=\sum_{i=0}^{n} f_{2i}\cos(Qx_{2i})-\tfrac{1}{2}[f_{2n}\cos(Qx_{2n})+f_0\cos(Qx_0)] \quad ,$$

$$S_{2n}=\sum_{i=0}^{n} f_{2i}\sin(Qx_{2i})-\tfrac{1}{2}[f_{2n}\sin(Qx_{2n})+f_0\sin(Qx_0)] \quad ,$$

$$C_{2n-1}=\sum_{i=1}^{n} f_{2i-1}\cos(Qx_{2i-1}) \quad ,$$

$$S_{2n-1}=\sum_{i=1}^{n} f_{2i-1}\sin(Qx_{2i-1}) \quad .$$

Imaginary part can be similarly handled. In order to apply these formulas, we need to tabulate the differential exchange potential $\Delta V_x(\underline{r})$ on a mesh within the unit cell. As noted earlier, $\Delta V_x(\underline{r})$ is numerically small throughout and is a quite smooth function. Thus a rather coarse grid is usually sufficient. This may still require computations over many thousands of $\underline{r}$ points. The computation time, however, can be significantly reduced by tabulating the spherical atomic charge density on a uniform logarithmic mesh. Total charge density at any point in the unit cell can then be obtained by summing the contribution due to each atom obtained by using a three-point uniform interpolation scheme. Hence,

$$V_x^R(\underline{G})=\Delta V_x^R(\underline{G})+\frac{\pi^{3/2}}{\Omega}\sum_{\mu}\cos\underline{G}\cdot\underline{\tau}_{\mu}\sum_i\frac{a_i^{\mu}}{(\beta_i)^{3/2}}e^{-G^2/4\beta_i^{\mu}} \quad ,$$

$$V_x^I(\underline{G}) = \Delta V_x^I(\underline{G}) + \frac{\pi}{\Omega}^{3/2} \sum_\mu \sin\underline{G}\cdot\underline{\tau}_\mu \sum_i \frac{a_i^\mu}{(\beta_i)^{3/2}} e^{-G^2/4\beta_i^\mu} \quad ,$$

where $\Delta V_x^{R,I}$ are obtained by three-dimensional Filon integration formula where the oscillatory part of the integrand was carefully separated out. The other part is analytic and presents no problems.

Appendix 3. D Embedding Formalism

It is usually not possible in practice to treat large clusters without additional simplifying assumptions, and thus one is never quite sure as to what effect the rest of the solid has on the cluster calculated properties. The idea of embedding appears to be quite useful in providing this connection. Though the embedding formalism has been described in various papers [3.22,105-110], the implementation [3.111-114] is still in its evolutionary stages. We briefly describe now the embedding formalism and mention some of the difficulties it represents.

We follow here the development [3.22,105-108] due to GRIMLEY and PISANI for deriving the general embedding equations. In ordinary cluster calculations we consider a cluster C which consists of an adsorbate A and a "small" part of the solid B (C=A∪B). We are here interested in formally investigating the effect of the rest

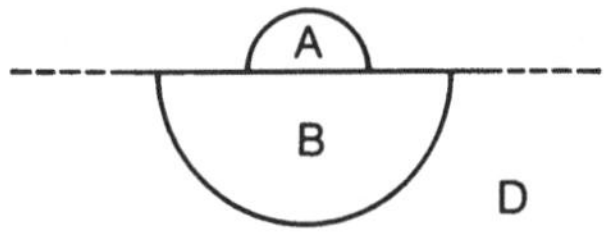

Fig. 3.14. Partitioning scheme for adsorbate-substrate for embedding purposes. The cluster (A∪B) is denoted by C in the text

of solid D ("the defective solid") on C (Fig.3.14). In the Green's function formalism, let G be the representation of the Green's operator in the basis set C ∪D, i.e.,

$$G^{-1} \equiv Q(\varepsilon) = (\varepsilon + i0)S - H \quad , \tag{3.85}$$

where H is the Hamiltonian matrix for the perturbed system and S the overlap matrix. By writing $Q(\varepsilon)G(\varepsilon) = I$ more explicitly:

$$\begin{pmatrix} Q_{CC} & Q_{CD} \\ Q_{DC} & Q_{DD} \end{pmatrix} \begin{pmatrix} G_{CC} & G_{CD} \\ G_{DC} & G_{DD} \end{pmatrix} = \begin{pmatrix} I_C & 0 \\ 0 & I_D \end{pmatrix} \quad , \tag{3.86}$$

one obtains for each energy ε

$$Q_{CC}G_{CC} + Q_{CD}G_{DC} = I_C \tag{3.87}$$

$$Q_{DC}G_{CC} + Q_{DD}G_{DC} = 0 \quad . \tag{3.88}$$

Substituting for G_{CC} from (3.88) into (3.87), one obtains

$$(G_{CC})^{-1} = Q_{CC} - Q_{CD}(Q_{DD})^{-1}Q_{DC} \quad . \tag{3.89}$$

If it is assumed that the perturbation potential V only affects the cluster region, i.e.,

$$V = H - H^{f} = Q^{f} - Q = \begin{pmatrix} V_{CC} & 0 \\ 0 & 0 \end{pmatrix}$$

(superscript f specifies "free solid" value), then

$$(G_{CC})^{-1} = Q_{CC} - Q^{f}_{CD}(Q^{f}_{DD})^{-1}Q_{DC} \quad . \tag{3.90}$$

Equation (3.90) may be put into a different form by subtracting the corresponding equations for the free solid,

$$G_{CC} = \left[(G^{f}_{CC})^{-1} - V_{CC}\right]^{-1} \quad . \tag{3.91}$$

Thus knowing the solution for the free solid G^{f}, the cluster part G^{f}_{CC} can be extracted. Then using (3.91), G_{CC} can be evaluated at each energy point. However, the problem must be solved self-consistently because the Hamiltonian matrix H[P] depends on the density matrix P_{CC}, which in turn is related to G_{CC}:

$$P_{CC} = -\frac{1}{\pi}\,\mathrm{Im}\int_{-\infty}^{\varepsilon^{f}_{F}} G_{CC}(\varepsilon)d\varepsilon \quad . \tag{3.92}$$

The main difficulty in this procedure lies in the necessity of inverting, at each step of the SCF state, relatively large matrices.

One can overcome this difficulty by following a slightly different procedure that requires approximation somewhat more stringent than the preceeding ones. To see this, let us develop (3.86) more explicitly in the cluster region C(AUB). Thus

$$Q_{AA}G_{AA} + Q_{AB}G_{BA} + Q_{AD}B_{DA} = I^{A} \quad ,$$

$$Q_{AA}G_{AB} + Q_{AB}G_{BB} + Q_{AD}G_{DB} = 0 \quad ,$$

$$Q_{BA}G_{AA} + Q_{BB}G_{BA} + Q_{BD}G_{DA} = 0 \quad ,$$

$$Q_{BA}G_{AB} + Q_{BB}G_{BB} + Q_{BD}G_{DB} = I^{B} \quad . \tag{3.93}$$

At this point three approximations are introduced. (1) Matrix elements of Q and G involving basis functions which belong to the defective solid are frozen at their "free solid" value, i.e., $Q_{BD} \simeq Q^{f}_{BD}$, $G_{DD} \simeq G^{f}_{DD}$, etc. (2) The Fermi level is kept fixed at the free solid value. (3) Compound quantities such as $Q_{AD}Q_{DB}$ or $Q_{AB}G_{BD}$

are neglected. While approximations (1) and (2) are implied in the former procedure, the (3) approximation is new and quite restrictive. It not only demands that the perturbation introduced by the adsorbate be negligible outside the region B, but also that it be adequately screened within B. In other words, this theory defines a solution to the problem which uses a local basis set but satisfies boundary conditions which are derived from the solution for the free solid. This can only be satisfied if the region B is sufficiently large. Thus the two main disadvantages of the present embedding techniques are that investigations can not really be restricted to small clusters. At the same time they require that the results for the periodic solid be known. For the free solid we have

$$Q^f G^f = I$$

from which we get

$$Q^f_{BB} G^f_{BB} + Q^f_{BD} G^f_{DB} = I^B \quad . \tag{3.94}$$

Using approximations (1) and (2) and (3.94) in (3.93) we get

$$Q_{AA} G_{AA} + Q_{AB} G_{BA} = I^A \quad ,$$

$$Q_{AA} G_{AB} + Q_{AB} G_{BB} = 0 \quad ,$$

$$Q_{BA} G_{AA} + Q_{BB} G_{BA} = 0 \quad ,$$

$$Q_{BA} G_{AB} + Q_{BB} G_{BB} = Q^f_{BB} G^f_{BB} \quad .$$

Or equivalently

$$Q_C G_C = J_C \quad , \tag{3.95}$$

where

$$J_C = \begin{bmatrix} I^A & 0 \\ 0 & Q^f_{BB} G^f_{BB} \end{bmatrix} \tag{3.96}$$

To solve the system of equations (3.95), multiply the left side by the inverse of Q_C (and call the inverse of Q_C an auxiliary matrix $\overline{G}_C$),

$$G_C(\varepsilon) = \overline{G}_C(\varepsilon) J_C(\varepsilon) \quad . \tag{3.97}$$

The advantage of this equation is that the problem is now efficiently factorized into a standard diagonalization within the cluster and into matrix multiplications. The matrix $\overline{G}_C$ can be constructed from the eigenvectors $\overline{A} = \{\overline{a}_{mi}\}$ and eigenvalues $\overline{E} = \{\overline{\varepsilon}_i \delta_{ij}\}$ of the cluster Hamiltonian:

$$H_C\bar{A} = S_C\bar{A}\bar{E} \quad ,$$

$$\bar{A}^+ S_C \bar{A} = I_C \quad .$$

Then

$$\bar{G}_{m\ell}(\varepsilon) = [Q_C(\varepsilon)]^{-1}_{m\ell} = \sum_i \bar{a}_{mi}\bar{a}_{\ell i}\left[P\{\frac{1}{\varepsilon-\varepsilon_i}\} - i\pi\delta(\varepsilon-\bar{\varepsilon}_i)\right] \quad .$$

Here P stands for "principal part of". Recalling that

$$P_C = -\frac{1}{\pi}\,\mathrm{Im}\{\int_{-\infty}^{\varepsilon_F^f} G_C(\varepsilon)d\varepsilon\} = -\frac{1}{\pi}\,\mathrm{Im}\{\int \bar{G}_C(\varepsilon)J_C(\varepsilon)d\varepsilon\} = -\frac{1}{\pi}\,\mathrm{Im}\{\int \bar{G}_C(\varepsilon)Q_C^f G_C^f d\varepsilon\} \quad ,$$

one can write

$$P_{mn} = \sum_{\ell,i} \bar{a}_{mi}\bar{a}_{\ell i}M_{\ell n}(\bar{\varepsilon}_i) \quad , \tag{3.98}$$

where the matrix M depends only on the electronic structure of the free solid and need to be computed only once. Thus the eigenvectors of the cluster allow the determination of the density matrix which in turn define the Hamiltonian H^C, and thus the problem can be solved iteratively. Knowing the Green's function for the system, one can calculate the single particle properties. However, as alluded to earlier, the above embedding development is beset with at least two serious computational difficulties. The first one is connected with the size of the local cluster, which has to be sufficiently large. The second is the requirement that the solution for bulk or the semiinfinite solid is assumed to be known which in practice is a major undertaking. However, once a bulk solution is known, various adsorbates can be placed on it without recomputing the bulk solid.

Another embedding technique is due to MUSCAT and MEWNS [3.115] who essentially embed their transition metal cluster in a jellium. A somewhat related approach is due to WHITTEN and PAKKANEN [3.116]. They represent the solid lattice by a large cluster and use the Hartree-Fock approximation using only valence electrons of the system. The local surface region of interest is created by a unitary localization transformation of the single-particle orbitals of the lattice wave function. From this set a partial list (with large localization) is selected and further augmented with additional basis functions on surface atoms to allow greater flexibility. Finally, the adsorbate and local surface region are treated (by configuration interaction at an ab initio level) as embedded in the fixed field of the lattice. This method appears quite promising but has an obvious drawback in that the lattice is still represented by a cluster which has to be large, but of still unknown size.

References

3.1 J. Koutecky: Adv. Chem. Phys. *9*, 85 (1965)
3.2 S.G. Davisson, J.D. Levine: Solid State Phys. *25*, 1 (1970)
3.3 J.R. Schrieffer: J. Vac. Sci. Technol. *9*, 561 (1971); *13*, 335 (1976)
3.4 K.H. Johnson: Adv. Quantum Chem. *7*, 143 (1973)
3.5 N.D. Lang: Solid State Phys. *28*, 225 (1973)
3.6 J.R. Smith: in *Interaction on Metal Surfaces*, ed. by R. Gomer, Topics in Applied Physics, Vol. 4 (Springer, Berlin, Heidelberg, New York 1975) p. 1
3.7 J.A. Appelbaum, D.R. Hamann: Rev. Mod. Phys. *48*, 479 (1976)
3.8 T.B. Grimley, in *Chemistry and Physics of Solid Surfaces*, ed. by R. Vanselow, S.Y. Tong (CRC, Ohio 1977) p. 155
3.9 G. Allan: in *Handbook of Surfaces and Interfaces Vol. 1*, ed. by L. Dobrzynski (Garland, New York 1978) p. 299;
R. Haydock, V. Heine, M.J. Kelly: J. Phys. C *5*, 2845 (1972);
L. Hodges, R.E. Watson, H. Ehrenreich: Phys. Rev. B *5*, 3953 (1972);
F. Cyrot-Lackmann, M.C. Desjonqueres: Surf. Sci. *40*, 423 (1973);
F. Yndurain, L.M. Falicov: J. Phys. C *8*, 1571 (1975)
3.10 C.W. Bauschlicher Jr., P.S. Bagus, H.F. Schaefer III: IBM J. Res. Develop. *22*, 213 (1978)
3.11 R.P. Messmer: in *The Nature of the Surface Chemical Bond*, ed. by T.N. Rhodin, G. Ertl (North Holland, Amsterdam 1979) p. 51;
T.N. Rhodin, J.W. Gadzuk: ibid., p. 113
3.12 M.L. Cohen: Physics Today *32*, 40 (1979); Nature *278*, 688 (1979);
M. Schlüter: J. Vac. Sci. Technol. *16*, 1331 (1979)
3.13 F. Herman: J. Vac. Sci. Technol. *16*, 1101 (1979)
3.14 W.A. Goddard III, T.C.McGill: J. Vac. Sci. Technol. *16*, 1308 (1979)
3.15 R.B. Laughlin, J.D. Joannopoulos, D.J. Chadi: J. Vac. Sci. Technol. *16*, 1327 (1979)
3.16 A. Zunger: J. Vac. Sci. Technol. *16*, 1337 (1979)
3.17 S.T. Pantelides, J. Pollman: J. Vac. Sci. Technol. *16*, 1349 (1979);
K.C. Pandey, J.C. Phillips: Phys. Rev. B *13*, 750 (1976);
D.J. Chadi: Phys. Rev. B *19*, 2074 (1979)
3.18 a) I.P. Batra: J. Vac. Sci. Technol. *16*, 1359 (1979);
b) C.S. Wang, A.J. Freeman: Phys. Rev. B *19*, 793 (1979);
c) A. Rosén, E.J. Baerends, D.E. Ellis: Surf. Sci. *82*, 139 (1979);
d) P.J. Feibelman, J.A. Appelbaum, D.R. Hamann: Phys. Rev. B *20*, 1433 (1979)
3.19 J.P. Muscat, D.M. Newns: Prog. Surf. Sci. *9*, 1 (1978)
3.20 T.L. Einstein, J.A. Hertz, J.R. Schrieffer: in *Theory of Chemisorption*, ed. by J.R. Smith. Topics in Current Physics, Vol. 19 (Springer, Berlin, Heidelberg, New York 1980) p. 183
3.21 W.A. Harrison: *Pseudopotentials in the Theory of Metals* (Benjamin, New York 1966);
M.L. Cohen, V. Heine: Solid State Phys. *24*, 37 (1970);
J.C. Phillips: *Bonds and Bands in Semiconductors* (Academic , New York 1973)
3.22 T.B. Grimley: in *Proceedings of the International School of Physics Enrico Fermi, Course LVIII*, ed. by F.D. Goodman (Compositori, Bologna 1974) p. 298
3.23 M.F. Thorpe, D. Weaire: Phys. Rev. Lett. *27*, 1581 (1971);
J.D. Jannopoulos, M.L. Cohen: Solid State Phys. *31*, 71 (1976)
3.24 K. Hirabayashi: J. Phys. Soc. Japan *27*, 1475 (1969);
G.P. Alldredge, L. Kleinman: Phys. Rev. Lett. *28*, 1265 (1972);
R.V. Kasowski: Phys. Rev. Lett. *33*, 83 (1974);
M. Schluter, J.R. Chelikowsky, S.G. Louie, M.L. Cohen: Phys. Rev. Lett. *34*, 1385 (1975);
W. Kohn: Phys. Rev. B *11*, 3756 (1975);
N. Kar, P. Soven: Phys. Rev. B *11*, 3761 (1975);
H. Krakauer, B.R. Cooper: Phys. Rev. B *16*, 605 (1977);
J.G. Gay, J.R. Smith, F.J. Arlinghaus: Phys. Rev. Lett. *38*, 561 (1977);
S. Ciraci, I.P. Batra: Phys. Rev. B *15*, 3254 (1977);
K. Mednic, C.C. Lin: Phys. Rev. B *17*, 4807 (1977);

O. Jepsen, J. Madsen, O.K. Andersen: Phys. Rev. B *18*, 605 (1978);
S.G. Louie: Phys. Rev. Lett. *40*, 1525 (1978);
H. Krakauer, M. Posternak, A.J. Freeman: Phys. Rev. B *19*, 1706 (1979)
3.25 M. Born, J.R. Oppenheimer: Ann. Phys. (Leipzig) *84*, 457 (1927)
3.26 J.C. Slater: *Quantum Theory of Matter* (McGraw-Hill, New York 1968)
3.27 D. Pines: *Elementary Excitations in Solids* (Benjamin, New York 1964)
3.28 T. Koopmans: Physica *1*, 104 (1933)
3.29 J.C. Slater: Phys. Rev. *81*, 385 (1951); *91*, 528 (1953);
P.A.M. Dirac: Proc. Cambridge Phil. Soc. *26*, 376 (1930)
3.30 W. Kohn, L.J. Sham: Phys. Rev. A *140*, 1113 (1965);
R. Gaspar: Acta Phys. Hung. *3*, 263 (1954);
L. Hedin, S. Lundqvist: Solid State Phys. *23*, 1 (1969);
A.K. Rajagopal: Adv. Chem. Phys., Vol. XLI, p. 59 (1980)
3.31 K. Schwarz: Phys. Rev. B *5*, 2466 (1972)
3.32 J.C. Slater: *Quantum Theory of Molecules and Solids*, Vol. IV, The Self-Consistent Field for Molecules and Solids (McGraw-Hill, New York 1974)
3.33 J.C. Slater: J. Chem. Phys. *43*, S228 (1965);
K.H. Johnson: J. Chem. Phys. *45*, 3085 (1966);
K.H. Johnson, F.C. Smith Jr.: Phys. Rev. B *5*, 831 (1972)
3.34 J. Korringa: Physics *13*, 392 (1947);
W. Kohn, N. Rostoker: Phys. Rev. *94*, 1411 (1954)
3.35 J.B. Danese, J.W.D. Connoly: J. Chem. Phys. *61*, 3063 (1974)
3.36 F. Herman, A.R. Williams, K.H. Johnson: J. Chem. Phys. *61*, 3508 (1974)
3.37 D.E. Ellis, G.S. Painter: Phys. Rev. B *2*, 2887 (1970)
3.38 D.J.M. Fassaert, H. Verbeek, A. van der Avoird: Surf. Sci. *29*, 501 (1972)
3.39 F. Bloch: Z. Phys. *52*, 555 (1928)
3.40 J.C. Slater, G.F. Koster: Phys. Rev. *94*, 1498 (1954)
3.41 This is a major field of research and we cannot possibly describe all the literature. The few references cited here are to be used as examples.
C.C.J. Roothann: Rev. Mod. Phys. *32*, 179 (1960);
E. Clementi: J. Chem. Phys. *38*, 996 (1963)
I. Schavitt: *Methods in Computational Physics*, Vol. 2 (Academic, New York 1963) p. 1
3.42 R.C. Chaney, T.K. Tung, C.C. Lin, E.E. Lafon: J. Chem. Phys. *52*, 361 (1970);
H. Sambe, R.M. Felton: J. Chem. Phys. *62*, 1122 (1975)
3.43 P.P. Ewald: Ann. Phys. (Leipzig) *64*, 253 (1921)
3.44 I.P. Batra, S. Ciraci, W.E. Rudge: Phys. Rev. B *15*, 5858 (1977)
3.45 S.F. Boys: Proc. Roy. Soc. (London) A *200*, 542 (1950)
3.46 S. Huzinaga: J. Chem. Phys. *42*, 1293 (1965)
3.47 H. Taketa, S. Huzinaga, K. O-Ohata: J. Phys. Soc. Japan *21*, 2313 (1966);
E. Clementi, J. Mehl: IBM Research Report RJ 883 (1971) (Unpublished)
3.48 N.D. Lang, A.R. Williams: Phys. Rev. Lett. *34*, 531 (1975); Phys. Rev. B. *18*, 616 (1978);
O. Gunnarsson, H. Hjelmberg, B.I. Lundquist: Phys. Rev. Lett. *37*, 292 (1976)
3.49 K.Y. Yu, J.N. Miller, P. Chye, W.E. Spicer, N.D. Lang, A.R. Williams: Phys. Rev. B *14*, 1446 (1976)
3.50 J. Harris, G.S. Painter: Phys. Rev. Lett. *36*, 151 (1976)
3.51 R.P. Messmer, D.R. Salahub: Phys. Rev. B *16*, 3415 (1977), Chem. Phys. Lett. *49*, 59 (1977); Int. J. Quantum Chem. *10S*, 183 (1976);
D.R. Salahub, R.P. Messmer: Phys. Rev. B *16*, 2526 (1977)
3.52 D.R. Salahub, M.R. Roche, R.P. Messmer: Phys. Rev. B *18*, 6495 (1978)
3.53 D.R. Salahub, Le-Chi Niem, M. Roche: Surf. Sci. *100*, 199 (1980)
3.54 R.P. Messmer, D.R. Salahub, J.W. Davenport: Chem. Phys. Lett. *57*, 29 (1978)
3.55 R.P. Messmer, C.W. Tucker Jr., K.H. Johnson: Surf. Sci. *42*, 341 (1974);
I.P. Batra, O. Robaux: J. Vac. Sci. Technol. *12*, 242 (1974); Surf. Sci. *49*, 653 (1975);
N. Rösch, D. Menzel: Chem. Phys. *13*, 243 (1976);
C.H. Li and J.W.D. Connoly: Surf. Sci. *65*, 700 (1977)
3.56 I.P. Batra, S. Ciraci: Phys. Rev. Lett. *39*, 774 (1977)

3.57 I.P. Batra, S. Ciraci: *Proc. 7th Int. Vacuum Cong. and 3rd Int. Conf. Solid Surfaces*, Vienna (1977), ed. by R. Dobrozensky, F. Rudenauer, F.P. Viehbock, A. Breth (F. Berger and Sohne, Vienna 1977) p. 1171
3.58 G.S. Painter: Phys. Rev. B *17*, 662 (1978)
3.59 D.W. Bullett: Surf. Sci. *93*, 213 (1980)
3.60 K. Kambe: Surf. Sci. (to be published)
3.61 E. Caruthers, L. Kleiman, G.P. Alldredge: Phys. Rev. B *8*, 4570 (1973); B *9*, 3325 (1974); B *9*, 3330 (1974);
J.R. Chelikowsky, M. Schlüter, S.G. Louie, M.L. Cohen: Solid State Commun. *17*, 1103 (1975); H.
H. Krakauer, M. Posternak, A.J. Freeman: Phys. Rev. Lett. *41*, 1072 (1978)
3.62 E.E. Huber Jr., C.T. Kirk Jr.: Surf. Sci. *5*, 447 (1966);
C.T. Kirk Jr., E.E. Huber Jr.: Surf. Sci. *9*, 217 (1968);
G. Dorey: Surf. Sci. *27*, 311 (1971);
W.H. Krueger, S.R. Pollack: Surf. Sci. *30*, 263 (1972)
3.63 L.H. Jenkins, M.F. Chung: Surf. Sci. *28*, 409 (1971)
3.64 P.O. Gartland: Surf. Sci. *62*, 183 (1977);
A.M. Bradshaw, P. Hofmann, W. Wyrobisch: Surf. Sci. *68*, 269 (1977)
3.65 K.L.I. Kobayashi, Y. Shiraki, Y. Katayama: Surf. Sci. *77*, 449 (1978)
3.66 P. Hofmann, W. Wyrobisch, A.M. Bradshaw: Surf. Sci. *80*, 344 (1979)
3.67 P.H. Dawson: Int. J. Mass Spectrum. Ion Phys. *16*, 269 (1975); Surf. Sci. *57*, 229 (1976)
3.68 B.J. Baird, C.S. Fadley: *Proc. Int. Study Conf. on Photoemission from Surfaces*, Noordwijk, Holland, ESA sp-118, ed. by R.F. Willis, B. Feuerbacher, B. Filton, C. Backx (ESA Scientific and Technical Publications, Noordwijk 1976) p. 173
3.69 R.J. Baird, C.S. Fadley, S.M. Goldberg, P.J. Feibelman, M. Šunjić: Surf. Sci. *72*, 495 (1978)
3.70 R.Z. Bachrach, A. Bianconi, S.A. Flodström: Ref. [3.57], p. 1205;
A. Bianconi, R.Z. Bachrach, S.A. Flodström: Solid State Commun. *24*, 539 (1977)
3.71 D.J. Fabian, J. Fuggle, L.M. Watson, A. Barrie, L. Latham: in *Band Structure Spectroscopy of Metals and Alloys*, ed. by D.J. Fabian, L.M. Watson (Academic, London 1973) p. 91
3.72 S.A. Flodström, L.-G. Petersson, S.B.M. Hagström: J. Vac. Sci. Technol. *13*, 280 (1976); Solid State Commun. *19*, 257 (1976)
3.73 S.A. Flodström, R.Z. Bachrach, R.S. Bauer, S.B.M. Hagström: Phys. Rev. Lett. *37*, 1282 (1976); Ref. [3.57], p. 869
3.74 C.W.B. Martinsson, L.-G. Petersson, S.A. Flodström, S.B. Hagström: [Ref. 3.68, p. 177]
3.75 A.M. Bradshaw, W. Domcke, L.S. Cederbaum: Phys. Rev. B *16*, 1480 (1977)
3.76 R.Z. Bachrach, S.A. Flodström, R.S. Bauer, S.B.M. Hagström, D.J. Chadi: J. Vac. Sci. Technol. *15*, 488 (1978); Ref. [3.57], p. 461
3.77 W. Eberhardt, C. Kunz: Surf. Sci. *75*, 709 (1978)
3.78 D. Norman, D.P. Woodruff: J. Vac. Sci. Technol. *15*, 1580 (1978)
3.79 S.A. Flodström, C.W.B. Martinson, R.Z. Bachrach, S.B. Hagström, R.S. Bauer: Phys. Rev. Lett. *40*, 907 (1978)
3.80 P. Hofmann, K. Horn, A.M. Bradshaw, K. Jacobi: Surf. Sci. *82*, L610 (1979)
3.81 F. Jona: J. Phys. Chem. Solids *28*, 2155 (1967)
3.82 R. Payling, J.A. Ramsay: in *5th Australian Vacuum Conf. Brisbane* (1976); p. 131: J. Phys. C *13*, 505 (1980)
3.83 C.W.B. Martinson, S.A. Flodström, J. Rundgren, P. Westrin: Surf. Sci. *89*, 102 (1979) and references therein
3.84 L.I. Johansson, J. Stöhr: Phys. Rev. Lett. *43*, 1882 (1979)
3.85 M.L. den Boer, T.L. Einstein, W.T. Elam, R.L. Park, L.D. Roelofs, G.E. Laramore: Phys. Rev. Lett. *44*, 496 (1980)
3.86 J. Grimbolt, J.M. Eldridge: Surf. Sci. (to be published);
A.M. Bradshaw, et al.: [Ref. 3.64]
3.87 N. Cabrera, J. Hamon: C.R. Acad. Sci. Paris *22H*, 1713 (1947)
3.88 E.A. Gulbransen, W.S. Wysong: J. Phys. Colloid. Chem. *51*, 1087 (1947)
3.89 G. Hass: J. Opt. Soc. Am. *39*, 532 (1949)

3.90 R.K. Hart: Proc. Royal Soc. (London) A *236*, 68 (1956)
3.91 M.S. Hunter, P. Fowle: J. Electrochem. Soc. *103*, 482 (1956)
3.92 P.E. Blackburn, E.A. Gulbransen: J. Electrochem. Soc. *107*, 944 (1960)
3.93 M.J. Dignam: J. Electrochem. Soc. *109*, 184 (1962); M.J. Dignam: J. Electrochem. Soc. *109*, 192 (1962)
3.94 M.J. Dignam, W.R. Fawcett, Böhni: J. Electrochem. Soc. *113*, 656 (1966)
3.95 J.J. Dignam, W.R. Fawcett: J. Electrochem. Soc. *113*, 663 (1966)
3.96 R.K. Hart, J.K. Maurin: Surf. Sci. *20*, 285 (1970)
3.97 C.W.B. Martinson, S.A. Flodström: Solid State Commun. *30*, 671 (1979)
3.98 W. Eberhardt, F.J. Himpsel: Phys. Rev. Lett. *42*, 1375 (1979)
3.99 P. Hofmann, C.V. Muschwitz, K. Horn, K. Jacobi, A.M. Bradshaw, K. Kambe, M. Scheffler: Surf. Sci. *89*, 327 (1979)
3.100 I.P. Batra: Phys. Rev. B *17*, 4114 (1978)
3.101 A. Liebsch: Phys. Rev. Lett. *38*, 248 (1977); Phys. Rev. B *17*, 1653 (1978)
3.102 C.S. Wang, A.J. Freeman: Phys. Rev. B *19*, 4930 (1979)
3.103 R.S. Mulliken: J. Chem. Phys. *33*, 1833 (1955)
3.104 Most of the data in this table has been obtained from D. Roy, J.D. Carette: in *Electron Spectroscopy for Surface Analysis*, ed. by H. Ibach, Topics in Current Physics, Vol. 4 (Springer, Berlin, Heidelberg, New York 1977) p. 13
3.105 T.B. Grimley: in *The Nature of the Surface Chemical Bond*, ed. by T.N. Rhodin, G. Ertl (North Holland, Amsterdam 1979) p. 1
3.106 T.B. Grimley: in *Electronic Structure and Reactivity of Metal Surfaces*, ed. by E.G. Derouane, A.A. Lucas (Plenum, New York 1976) p. 113
3.107 T.B. Grimley, C. Pisani: J. Phys. C *7*, 2831 (1974)
3.108 C. Pisani: Phys. Rev. B *17*, 3143 (1978)
3.109 J. Bernhole, S.T. Pantelides: Phys. Rev. B *18*, 1780 (1978)
3.110 O. Gunnarsson, H. Hjelmberg: Physica Scripta *11*, 97 (1975)
3.111 O. Gunnarsson, H. Hjelmberg, B.I. Lundqvist: Surf. Sci. *348* (1977)
3.112 R.A. Van Santen, L.H. Toneman: Int. J. Quantum Chem. *12*, Suppl. *2*, 83 (1977)
3.113 C. Pisani, R. Dovesi, P. Carosso: Phys. Rev. B *20*, 5345 (1979)
3.114 A. van der Avoird, S.P. Liebmann, D.J.M. Fassaert: Phys. Rev. B *10*, 1230 (1974)
3.115 J.P. Muscat, D.M. Newns: Surf. Sci. *87*, 643 (1979)
3.116 J.L. Whitten, T.A. Pakkanen: Phys. Rev. B *21*, 4357 (1980); T.H. Upton, W.A. Goddard III, C.F. Melius: J. Vac. Sci. Technol. *16*, 531 (1979)
3.117 H.L. Yu, M.C. Munoz, F. Soria: Surf. Sci. *94*, L184 (1980)
3.118 Ding-sheng Wang, A.J. Freeman, H. Krakauer: Phys. Rev. *B24*, 3092 (1981); *B24*, 3104 (1981)
3.119 B.N. Cox, C.W. Bauschlicher Jr.: Surf. Sci. (in press)
3.120 R.Z. Bachrach, G.V. Hansson, R.S. Bauer: Surf. Sci. *109*, L560 (1981).
3.121 L. Kleinman: (private communication). At the time of this writing the calculation was in progress

4. Computer Experiments on Heterogeneous Systems

B. J. Berne and R. V. Mikkilineni*

With 13 Figures

Since the advent of high-speed digital computers, computer simulation studies have contributed a great deal to the understanding of condensed matter. They have provided essential information not otherwise obtainable by present experimental techniques and have helped advance our theoretical understanding of the structure and dynamics of condensed matter [4.1]. In spite of these advances, it is only recently that these techniques have been used to probe heterogeneous systems containing interfaces between different phases. One difficulty has been the relatively small number of particles in an interface that contribute to the thermodynamic properties of relevance. Often these contributions are masked by the so-called bulk properties, arising from particles that are not in the interface. In recent years, many attempts have been made to develop techniques that overcome this difficulty and allow one to study equilibrium and nonequilibrium properties of interfaces.

The purpose of this chapter is to discuss some of the techniques that are currently being used to study the nature of the liquid-vapor interface, especially those techniques relevant to studies of the structure and dynamics of microclusters. No attempt is made to present a comprehensive review, but the list of references should provide a good source for further reading. Although the illustrations are heavily drawn form our own work, the references provide a good source of other examples. Our discussion will be limited to purely classical fluid interfaces.

The rest of the chapter is organized as follows. Section 4.1 focuses on methodology. Section 4.2 focuses on the liquid-vapor interface and the formation of droplets using realistic model potentials. Section 4.3 gives a few examples of idealized systems that have been explored. These studies provide the reference systems for theories on inhomogeneous systems.

4.1 Methodology

Molecular dynamics, the Monte Carlo method and stochastic dynamics are the three major computer techniques used to simulate condensed systems. In molecular dynamics,

*Previously known as M. RAO

Hamilton's equations of motion are integrated by finite difference techniques and the phase space orbit is determined [4.2]. The various properties of the system are then determined by time averages over the phase space orbits. The Monte Carlo method on the other hand, involves the stochastic generation of a sequence of configurations [4.3,4]. The stochastic process is so defined that the distribution of configurational states thus generated is the "equilibrium distribution function." Equilibrium averages are then determined by averaging the properties of interest over the configurational states. Stochastic molecular dynamics lies somewhere between full molecular dynamics [4.5] and the Monte Carlo method. Stochastic molecular dynamics is used to stimulate a subsystem in contact with a "bath." The equations of motion, of the particles in the subsystem, are integrated subject to stochastic forces due to the bath. Recent developments in this area make possible the study of the scattering and thermal accomodation of gas particles on solid surfaces. In the Monte Carlo method and molecular dynamics [4.6-9], there are practical limitations on the size of the systems than can be simulated. In certain applications, stochastic molecular dynamics can be engineered to simulate truly infinite systems—nevertheless, the results depend on the stochastic model adopted. In applications concerning microclusters, the systems are finite and may thus be studied using full molecular dynamics or Monte Carlo procedures. It is nevertheless interesting to observe that simulations of finite systems consisting of a few hundred to a few thousand particles with short-range interactions give excellent agreement with the thermodynamic properties of uniform cryogenic liquids over the whole liquid-gas coexistence curve [4.10].

All three simulation techniques can be devised to simulate systems in the microcanonical, canonical, grand canonoical, and isobaric ensembles. In the thermodynamic limit $N \to \infty$, $V \to \infty$, N/V = constant, these various ensembles give the same thermodynamic equations of state, but for finite systems, there will be differences between these ensembles. In simulating microclusters, it is important to pick the ensemble that is dictated by the particular experiment.

In most applications discussed here the system consists of a small number of particles, N, in a box of volume V. We are compelled to choose boundary conditions consistent with the physical system of interest. The results of the simulation will be sensitive to these boundary conditions and to the potential energy of interaction of the particles, that is, the assumed interparticle forces. Periodic boundary conditions are employed in the simulation of one phase uniform systems. Other types of boundary conditions are employed to simulate nonuniform systems with interfaces. These will be described later. The interaction potential is not generally known; thus idealized potential models are often adopted (e.g., hard-sphere fluid, square-well fluid, lattice-gas fluid, Lennard-Jones fluid, ST-2 water, overlap ellipsoidal molecules). These idealized interaction models have provided a wealth of information that has helped in the formulation of many theories and has led to the understanding of fluid interfaces with more realistic potentials.

Molecular dynamics simulations have the advantage that they can give both the equilibrium and nonequilibrium properties of a system. Monte Carlo simulations, on the other hand, give only the equilibrium properties. Stochastic dynamics combines the elements of the above two methods and facilitates the study of complex processes including chemical reactions. All these methods are discussed in great detail in the references [4.1-10], and we shall present only a brief discussion here.

4.1.1 Molecular Dynamics

In the past decade, significant advances have been made in our understanding of the structure and dynamics of simple fluids containing spherical particles. Much of the success in this field springs directly from molecular-dynamics calculations in which the equations of motion of a system containing a large number of particles are solved using a computer and the resulting phase space trajectory is used to calculate time averages of dynamical variables. These computer experiments, when coupled with analytical advances in theory, have essentially given rise to a renaissance in the study of fluids.

The literature on molecular dynamics had been growing rapidly. Rather than review here many of the important applications, we refer the reader to comprehensive reviews of the material [4.1,2].

According to the basic formalism of statistical mechanics, the equilibrium mechanical properties of a many-body system are given by time averages of dynamical variables. For example, the pressure is simply related to the time average of the virial of Clausius. In addition, according to linear response theory, transport coefficients and spectral line shapes are directly determined by time correlation functions which represent the way spontaneous fluctuations arise and regress in equilibrium systems [4.11].

Electronic computers have advanced to such a degree that Newton's equations of motion can now be integrated for a fairly large assembly of interacting particles. This method, now commonly called molecular dynamics, is particularly useful for the study of the condensed phases of matter. Features common to most molecular dynamics experiments to date are the following.

1) The systems are finite, $N \leq 4000$.

2) The interaction potential is pairwise additive

$$V(1,\dots,N) = \sum_{i<j=1}^{N} \phi(r_{ij}) \tag{4.1}$$

where the pair potential $\phi(r)$ has a finite range r_0 such that for $r > r_0$, $\phi(r) = 0$.

3) For continuous pair potentials, Newton's equations of motion

$$m \frac{d\underset{\sim}{v}_i}{dt} = - \sum_{j=1} \frac{\partial}{\partial \underset{\sim}{r}_i} \phi(r_{ij}); \; \underset{\sim}{v}_i = \frac{d\underset{\sim}{r}_i}{dt} \tag{4.2}$$

are solved by finite difference techniques with time steps Δt between 10^{-14} and 10^{-15} s; and the positions, velocities, and accelerations are stored on magnetic tape or disc.

4) For discontinuous potentials, like the hard-sphere potential, the instanteous positions and momenta of the particles allow one to calculate a table of collision times. The particles are advanced along free particle (linear) trajectories until the "next" binary collision occurs. At the instant of collision, the new momenta of the colliding pair are assigned, and the system is moved along free particle trajectories until the next collision, and so on.

5) In addition, computer experiments on equilibrium liquids usually have the following features in common.

a) The N particles are contained in a rectangular box (L_x, L_y, L_z) in which the shortest edge is $\geq 2r_0$ such that N and V ($=L_xL_yL_z$) are chosen to give the desired number density.

b) The initial state (initial positions and momenta) are sampled such that the momenta are distributed according to the Maxwell distribution at a given temperature and the positions are sampled such that the initial configuration corresponds to quite large values of the Boltzman factor $\exp-\{\beta V(1,\ldots N)\}$; i.e., the system is in a probable state for a member of an equilibrium ensemble. Sampling of the initial configuration can be generated by the Monte Carlo method or by assuming an initial lattice, followed by appropriate aging.

c) The equations of motion are solved subject to the initial state together with periodic boundary conditions: the latter means that if $\underset{\sim}{r}_i \equiv (x_i, y_i, z_i)$ is the position of particle i in the box, there are 26 periodic images at $\underset{\sim}{r}_i + (k_xL_x, k_yL_y, k_zL_z)$ where k_i assumes the values (-1, 0,1) and not all k's can be zero. The particles in the box interact with the periodic images within their range. Consequently, when a particle leaves the box through one side, its image enters through the opposite side, which thus preserves the number of particles in the box. These conditions eliminate strong surface effects and essentially simulate an infinite system. Nevertheless, periodic boundary conditions have certain limitations—especially where long-range forces are concerned.

d) The computer output is then the sequence of states $\Gamma_0,\ldots,\Gamma_j,\ldots,\Gamma_M$ through which the system passes in the course of time. Here Γ_j is the state of the system (all positions and momenta) after the jth time step, i.e., at time $t = j\Delta t$.

The output is consequently a set of discrete points in phase space and M is the total number of iterations done in the computations. Since the initial state might be an improbable state, it is customary to "age" the system to "equilibrium." The reader should consult the literature for details [4.1]. After a short aging run, the subsequent states can be regarded as typical of an equilibrium system at the total energy E, N and V. Henceforth, by Γ_0 will be meant the initial state of the equilibrated system.

e) The computer generates the state $(\Gamma_0,\ldots,\Gamma_j,\ldots,\Gamma_M)$ through which the system passes in the course of time so that any property $A(\Gamma_t)$ can be determined for each of these states giving $(A_0,\ldots,A_j,\ldots A_M)$.

Given the trajectory, both the time average $\langle A\rangle$ and autocorrelation function $\langle A(0)A(t)\rangle$ can be determined as follows:

$$\langle A\rangle = \frac{1}{M}\sum_{j=0}^{M} A_j \tag{4.3}$$

$$\langle A(0)A(\tau)\rangle = \frac{1}{M-n}\sum_{j=0}^{M-n} A_j A_{j+m}; \quad \tau = m\Delta t \quad , \tag{4.4}$$

f) To simulate a system, it is necessary to know the intermolecular potential. Unfortunately, accurate pair potentials are not available and it is often necessary to use model potentials. The most common potentials used for monatomic fluids are the smooth hard-sphere potential or the Lennard-Jones (6,12) (L.J.) potential. Extensive molecular dynamics calculations on "Lennard-Jonesium"—the model fluid in which the particles interact through the L.J. (6,12) potential—show that this model gives accurate thermodynamic and structural information over a wide range of fluid states [4,10]. The situation is much more complex for the interaction of polyatomic molecules. Considerable work has been done on potential models of water. We note in this regard the electrostatic model of STILLINGER et al. [4.11].

As mentioned before, highly idealized potential models such as the hard-sphere potential and the square-well potential have played a profound role in helping to shape our ideas about fluids [4.1]. These models have the advantage of requiring less computer time.

In a dense fluid, a given molecule interacts on the average simultaneously with N other molecules. Clearly, N increases with the range of the pair potential. For small systems, the cycle time of the calculation then varies as N^2, whereas for large systems, various bookkeeping devices [4.12] can reduce the cycle time dependence to $N \ln N$. Truncated potentials are often used to reduce n. As we shall see, this can have a dramatic effect on the results obtained for fluids with long-range interactions.

4.1.2 Stochastic Molecular Dynamics

There are many problems that would require so much computer time that their study by molecular dynamics would not be possible. For example, the coagulation of proteins or colloids in various solvents involves such a large number of solvent and solute degrees of freedom that a full molecular dynamics simulation is unthinkable. The study of these systems requires an entirely different approach. Because we are only interested in the motion of the colloid particles, it is possible, using projection operator techniques, to project out the colloid motion [4.13]. If $(\underset{\sim}{R}_i, \underset{\sim}{V}_i)$ denote the position and velocity of colloid particle j, it can be shown that

$$\underset{\sim}{V}_i(t) \; E \; \dot{\underset{\sim}{R}}_i(t) \quad , \tag{4.5}$$

$$M_i \dot{\underset{\sim}{V}}_i(t) = \sum_j \underset{\sim}{F}_{ij} \left[|\underset{\sim}{R}_i(t) - \underset{\sim}{R}_j(t)| \right] - \int_0^t dt' K_i(t-t') \underset{\sim}{V}_i(t') + \underset{\sim}{f}_i(t) \quad ,$$

where $\underset{\sim}{F}_{ij}$ is the direct force on colloid particle i due to colloid particle j at time t, $\underset{\sim}{f}_i(t)$ is the fluctuating force acting on i due to the solvent, and $K_i(t)$ is the memory function representing the frictional effect of the solvent on particle j. By the second fluctuation-dissipation theorem [4.13],

$$K_j(t) = \frac{\langle f_j(0) f_j(t) \rangle}{\langle v_j^2 \rangle} \tag{4.6}$$

the memory function is related to the autocorrelation function of the fluctuating force.

To proceed it is necessary to adopt a stochastic model for the fluctuating force. For example, when the solvent or bath motion is fast compared to the system motion, $\underset{\sim}{f}_i$ fluctuates rapidly, $K_i(t)$ decays rapidly, and the equations reduce to ordinary Langevin equations. Usually, $\underset{\sim}{f}_i$ is assumed to be a Gaussian random process, with an assumed covariance $K_j(t)$. Various sampling techniques exist for sampling such a process and for solving the resulting generalized Langevin equations. This must be done subject to whatever boundary conditions are imposed on the system [4.9].

One area in which these techniques have been exploited concerns the scattering of gas particles from solid surfaces [4.6,14]. There, a gas is assumed to interact directly with a small patch of the solid. The particles in this patch are coupled to a bath consisting of the rest of the solid. The memory kernels are determined for the harmonic solid and used in the generalized Langevin equations. In this way the problem reduces to a very tractable set of equations. This approach gives accurate information about sticking probabilities, energy transfer, chemical reactions on surfaces, etc., and should be very useful in studies of aerosol systems.

4.1.3 Monte Carlo Method

In the Monte Carlo (MC) method [4.3,8] various configurations of a fluid are sampled according to the Boltzmann distribution

$$Z^{-1}e^{-\beta V(\underset{\sim}{R}_1,\ldots,\underset{\sim}{R}_N)}dr_1\ldots d\underset{\sim}{R}_N \quad , \tag{4.7}$$

where R_j is a vector specifiying both the position and orientation of molecule j, and Z is the canonical partition function. To generate a set of configurations that are so distributed, METROPOLIS et al. [4.3] devised a scheme according to which a new configuration $\underset{\sim}{R}' = (R_1',\ldots,R_N')$ is generated from an old configuration $\underset{\sim}{R} = (\underset{\sim}{R}_1,\ldots,\underset{\sim}{R}_N)$ by sampling a prescribed transition probability $T(\underset{\sim}{R}'|\underset{\sim}{R})$. This new configuration is accepted with probability

$$p = \mathrm{Min}\left[1, \frac{T(R|R')e^{-\beta V(R')}}{T(R'|R)e^{-\beta V(R)}}\right] \tag{4.8}$$

and rejected with probability $q = 1 - p$. In this way, a random walk over configuration space is generated, and it can be shown that after a sufficient number of moves are made, the configurations are distributed according to the Boltzmann distribution. The efficiency of this method depends on how rapidly it covers the available configuration space. More will be said about this later. Averages over the Boltzmann distribution are then obtained by weighting old configurations with q and new configurations with p. In this way, thermodynamic properties are predicted. The method has been frequently applied to the study of the structure of fluids, where an average over the sampled configuration gives the pair correlation functions.

Usually the system is subjected to periodic boundary conditions (as in molecular dynamics). The configuration is changed by single molecule moves. First, particle 1 is moved and this move is accepted or rejected according to p or q, then this is repeated for molecules 3, 4, ..., N. The full cycle of moves is then repeated. Thus all that needs be specified is the transition probability for a single molecule move. For an atomic fluid the move is chosen uniformly so that

$$T(\underset{\sim}{R}_j'|\underset{\sim}{R}_j) = \begin{cases} (\Delta x_0 \Delta y_0 \Delta z_0)^{-1}, & \delta R_j \varepsilon D \\ 0, & \delta R_j \varepsilon D \end{cases} \quad , \tag{4.9}$$

where D defines a domain of possible moves $(\frac{-\Delta x_0}{2} \le x \le \frac{\Delta x_0}{2}$, etc.). Similarly, for molecules, the translational and rotational moves are sampled randomly from some domain D. For this kind of uniform sampling, (4.8) reduces to

$$p = \mathrm{Min}\left[1, \frac{e^{-\beta V(R')}}{e^{-\beta V(R)}}\right] \quad . \tag{4.10}$$

We call uniform sampling the Metropolis scheme.

In many systems such as fluids with hydrogen bonds, there can be configurational bottlenecks. It has been found for these systems that molecular dynamics converges more rapidly than the Metropolis scheme.

To compare these methods, BERNE et al. [4.15] have studied the translational and rotational diffusivity in molecular dynamics and in the Metropolis walk. They find that the molecular dynamics (MD) method covers configuration space more quickly and more efficiently than the Metropolis walk. To achieve rapid convergence in the MC method, BERNE et al. have devised a method for Monte Carlo sampling which they call the Force Bias scheme [4.15]. In this scheme, the transition probability is taken as

$$T(\underset{\sim}{R}_j{}', \underset{\sim}{R}_j) = \begin{cases} c e^{-\beta/2\, [\nabla_{\underset{\sim}{R}_j} V(R)] \cdot \delta R_j}, & \delta R_j \in D \\ 0 \quad , & \text{otherwise,} \end{cases} \tag{4.11}$$

and movements in the direction of the force are sampled with greater step sizes than movements in other directions. Compared with the Metropolis scheme given by (4.8,9), it is clear that this new scheme samples moves more in keeping with the forces and torques on a molecule, as does molecular dynamics. This scheme generates a walk with greater diffusivity than does the Metropolis walk and is useful in treating aqueous systems. The Force Bias scheme arises quite naturally out of stochastic dynamics. Other optimization methods have been discussed by OWICKI and SCHERAGA [4.16].

The Monte Carlo method, its successes and its limitations, have been reviewed extensively in the literature [4.4,17]. The original method has been modified to improve its efficiency [4.15,18] and extended to simulate other types of ensembles (grand canonical, isobaric, etc.).

4.2 Computer Simulation of Planar Interfaces in a Lennard-Jones Fluid

To study a realistic system in a computer simulation, the interaction potential must be chosen carefully. It is well known that the bulk properties of fluids at liquid densities depend very much on the short-range repulsive part of the interaction potential. This is the reason for the success of Van-der-Walls-like theories [4.2] in predicting the bulk properties of liquids. On the other hand, the interfacial properties are sensitive to the long-range attractive part of the interaction potential. While the interaction potential must, therefore, be chosen very carefully to mimic real interfaces, these studies also provide a stringent test of the potential. Thus, by comparing the simulation results with experiments in a laboratory, insight can be obtained on the nature of the potential.

The Lennard-Jones (6,12) potential has been widely adopted in studies of the bulk properties of liquids [4.2]. In recent years, this potential has been widely employed to study the two phase interface and thermodynamics. The potential of interaction between two particles i and j separated by a distance r is given by

$$V(r) = 4\varepsilon\left[\left(\frac{\sigma}{r}\right)^{12} - \left(\frac{\sigma}{r}\right)^{6}\right] \quad . \tag{4.12}$$

With $\sigma = 3.405$ Å and $\varepsilon = 119.4$ K, this potential represents bulk liquid argon closely [4.2]. It is common to express all distances in units of σ and all energies in units of ε. Then the reduced temperature is $\hat{T} = kT/\varepsilon$, the reduced pressure is $\hat{P} = P\sigma^3$.

In the rest of this section, we shall present results obtained based on this potential and elucidate our current understanding of the liquid-vapor interface derived from these studies.

4.2.1 Flat Interfaces

The simplest of the liquid-vapor interfaces is a flat interface. The structure of a planar interface has received much theoretical attention [4.19]. In such an interface, the density profile $n(\underset{\sim}{r})$ will be a function of a coordinate (taken as the z coordinate) in the direction perpendicular to the surface. $n(r)$ will be a constant, given by the liquid density in the region where there is bulk liquid, and by the gas density in the region where there is homogeneous gas phase. Thus, the inhomogeneity of the density profile is restricted to the region of the transition zone between liquid and vapor.

Another distribution function that characterizes the flat interface is the two-particle distribution $\rho_2(\underset{\sim}{r}_1,\underset{\sim}{r}_2)$. This function gives the number of distinct pairs of particles at $\underset{\sim}{r}_1$ and a $\underset{\sim}{r}_2$ per unit volume in pair space. Because of symmetry, in a flat interface, $\rho_2(\underset{\sim}{r}_1,\underset{\sim}{r}_2)$ is only a function of the heights z_1, z_2 and

$$u_{12} = \sqrt{x_{12}^2 + y_{12}^2} \quad , \tag{4.13}$$

where $\underset{\sim}{u}_{12}$ is a vector parallel to the plane of the interface. Most of the thermodynamic functions like the surface tension and the pressure tensor are described in terms of these distribution functions. Thus, any determination of these functions from computer simulation allows one to calculate the thermodynamic properties of the liquid-vapor interface.

a) *Preparation of the Planar Interface*

Many different methods have been employed to prepare an initial configuration that mimics the asymmetry of a physical planar interface. In most cases, a sample is confined in a box having the form of a square prism extending through the interface. In the two directions parallel to the surface, ordinary periodic boundary conditions are applied. In the third direction perpendicular to the interface, a boundary conditions is chosen to allow for the interface. Various methods have been employed. LEE et al. [4.20] employed a weak external potential that simulates gravity and stabilizes the liquid-vapor interface. LIU et al. [4.21] employed an external wall potential. RAO and LEVESQUE [4.22] used a self-bound liquid argon film in equilibrium with its own vapor. To create this film, they first equilibrated a homogeneous Lennard Jones fluid with periodic boundary conditions using a rectangular cell L×L in the x-y plane, and length M along the z direction. The primary cell was then removed and placed in the center of a rectangular box of cross-sectional area L×L, but of thickness M surrounded by a vacuum of thickness M on both sides. Subsequent simulation led to evaporation and finally to sheets of liquid in equilibrium with vapor (Fig.4.1). The specific geometry established is the key in maintaining the planar interface. In free space, without imposed gravity, a volume of liquid will ultimately take on the minimum surface-volume shape of a sphere to within the fluctuations and curvature effects. In a domain with two periodic boundary conditions, only two of the six surfaces which would usually bound a rectangular box will appear as free boundaries. It is expected that a planar sheet of thickness h will still be stable against spherical droplet formation if its volume to surface ratio exceeds that of a sphere, that is, if

$$h > (2/9\pi)^{\frac{1}{2}} L = 0.27L \quad .$$

When this condition is satisfied, a film in equilibrium with its own vapor should be stable.

It is now established that the initial equilibration can be expedited by sprinkling a few particles uniformly (in a random fashion) in the vacuum region on either side of the liquid film. The closer the density of this vapor to the correct vapor density at the given temperature, the faster will be the approach to equilibrium.

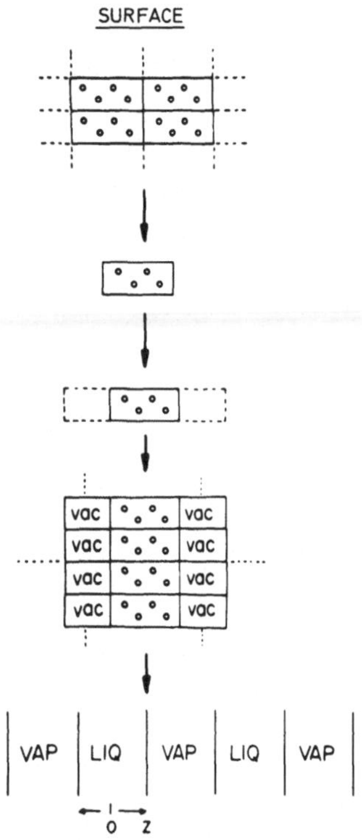

Fig. 4.1. Schematic drawing of the system used to simulate liquid-vapor free surface

b) *Density Profile*

Molecular dynamics and the Monte Carlo method have been used to study the density profile of a liquid-vapor interface in great detail. CROXTON and FERRIER, [4.23], using a two-dimensional Lennard-Jones fluid (the interface is just a line), obtained an oscillatory density profile in the interface region. The MC results of LEE et al. [4.20] also suggested a layering of the profile in the interfacial region of a three-dimensional Lennard-Jones fluid. Similar layering was observed by LIU [4.22].

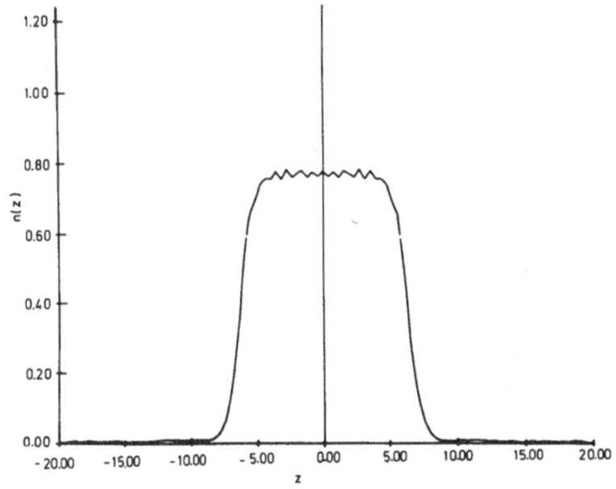

Fig. 4.2. Surface density profile $n(z)$ for a L.J. fluid at 84 K averaged over 6400 configurations

Using the self-bound liquid argon film, described earlier, RAO and LEVESQUE [4.22] carried out a very long molecular dynamics run (3900 time steps, corresponding to 3.9×10^{-10} sec.) and showed that the density profile is smooth and a monotone function going from the liquid density in the bulk phase to the gas density in the vapor phase (Fig.4.2).

ABRAHAM et al. [4.24], using a Monte Carlo simulation, also observed the monotone density profile, and their studies clearly showed that the layering observed in earlier MC results was due to insufficient statistics. Unfortunately, the density profile converges very slowly and extremely long simulations are required to obtain accurate results. In addition, there is an extra complication when an external wall potential is used to stabilize the liquid film. The wall potential usually introduces extra correlations manifesting themselves in an oscillatory profile. These oscillations are real and will not be averaged out even in an infinitely long simulation. Such oscillatory profiles have also thrown much light on the liquid-vapor interface [4.25].

The smooth monotone density profile suggests [4.22] that a local thermodynamic picture for the inhomogeneous fluid mimics, on a microscopic scale, conceptual aspects of the bulk phase. Using this picture, interfacial properties can be predicted from a knowledge of the bulk properties of the fluid. Following these lines, various recipes [4.26] have been proposed to explain the density profile. These arguments, while successful in explaining the longitudinal correlations, are quite inadequate to represent accurately what happens in the interface when one considers the transverse correlations [4.27].

c) *Transverse Correlations*

We have seen that in a planar interface, the two-particle distribution function is a function of z_1, z_2, and the transverse coordinate u_{12}. We shall define a two-dimensional Fourier transform of the two-particle distribution function [4.25],

$$S_T(k,z) = \frac{\sum_{i=1}^{N(\delta z)} \sum_{i=1}^{N(\delta z)} \exp(-i\underset{\sim}{k}\cdot\underset{\sim}{r}_{ij})}{\langle N(\delta z)\rangle} \tag{4.14}$$

where $\underset{\sim}{k}$ is a wave vector in the interfacial plane, the sum is over all particles in a sheet of thickness δz parallel to the surface, and the angular brackets denote the average over all configurations and an average over all value of k^2 within a small shell. $S_T(k,z)$ is shown in Fig.4.3 as a function of k in the case where z is situated in the interfacial region of a system containing a liquid film in equilibrium with its own vapor. The system contained 1728 particles at 84 K in a box of size $13.15 \times 13.15 \times 49.45\ \sigma^3$. The circles denote the results obtained using 1024 particles in a box of size $6.78 \times 6.78 \times 56.24\ \sigma^3$.

There is an apparent divergence as $k \to 0$, indicating transverse correlations over the size of the box. The genesis of the singular low k transverse behavior has been suggested by WERTHEIM [4.28] and is discussed in detail in relation to the computer simulation results by KALOS et al. [4.29]. This analysis shows that the singular low k transverse behavior is due to the capillary waves thermally excited in the surface.

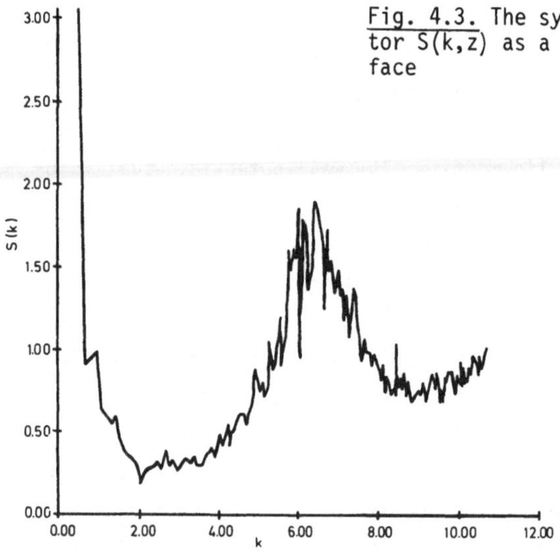

Fig. 4.3. The symmetrized transverse structure factor S(k,z) as a function of k where z is in the interface

The implication of this analysis is that gas and liquid are predominantly present only in very large clusters, so that in the course of time, a given point - or a set of points - can be regarded as either belonging to the liquid cluster or gas cluster. Thus the two-phase system consists of a temporal sequence or a phase space ensemble of configurations each sharply divided into gas and liquid regions. The instantaneous structure (Fig.4.4) is similar to that of a vibrationally excited membrane of drum head with some intrinsic thickness. The thermally excited waves - capillary

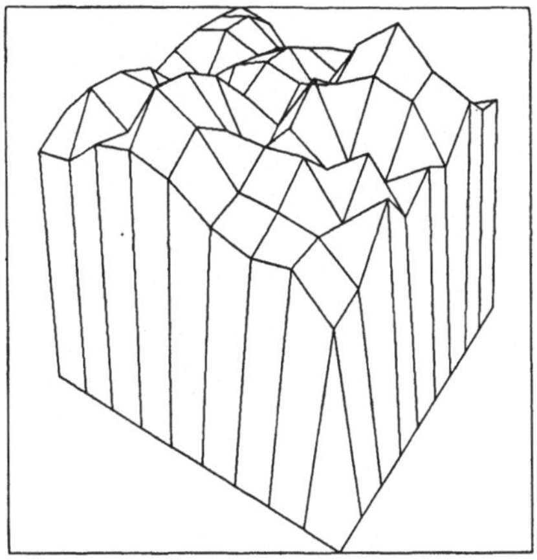

Fig. 4.4. Typical surface configuration of atoms with energy criterion $E_0 = -0.5$

waves - arise and decay like any thermal fluctuations in the liquid phase. The intrinsic thickness of the interface Δz is less than the intermolecular spacing in the liquid phase near the triple point.

The observed relatively broad density profile is then the average of an ensemble of gas-liquid boundaries in the form of the thermally excited capillary waves. Using this picture, RAO et al. [4.25] have shown that a liquid-vapor interface is perturbed by a hard wall in the interfacial region. The wall damps out the capillarity waves and induces an oscillating density profile (Fig.4.5). Simple capillary wave theory predicts these observations.

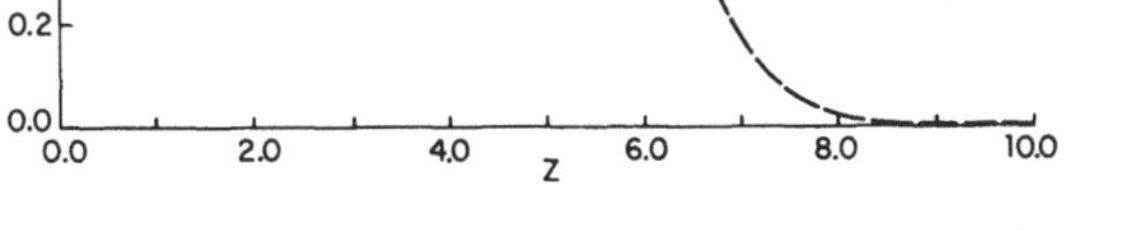

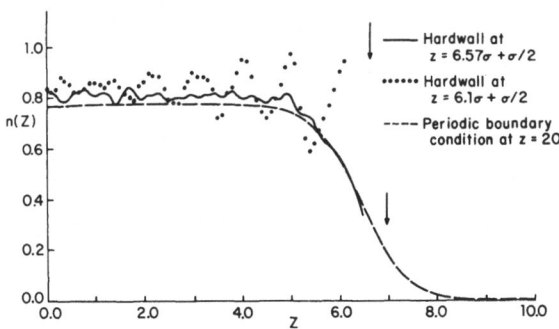

Fig. 4.5. The symmetrized density profile n(z) as a function of z. The dashed lines show the free interface (from [4.8]; the full line shows the density profile when walls are located at ±7.07; the dots show the density profile when the walls are located at ±6.6σ

d) *Pressure Tensor*

A knowledge of the two-particle distribution function $\rho_2(r_1,r_2)$ also yields information about the pressure tensor. The statistical mechanical expression relating $\rho_2(r_1,r_2)$ to the pressure tensor has been discussed by BUFF [4.30] and ONO and KONDO [4.31]. Following reference [4.31], the pressure tensor in a system with surface asymmetry can be written as

$$\underset{\approx}{P}(\underset{\sim}{r}) = n(\underset{\sim}{r})k_BT - \frac{1}{2}\langle\sum_{i\neq j}^{N}\int_{\infty}^{\infty} d\eta\,\theta(\eta)\theta(1-\eta)\frac{\underset{\sim}{r}_{ij}\underset{\sim}{r}_{ij}}{|\underset{\sim}{r}_{ij}|}\phi'(\underset{\sim}{r}_{ij})\delta[\underset{\sim}{r}-(\underset{\sim}{r}_i+\eta\underset{\sim}{r}_{ij})]\rangle \quad , \qquad (4.15)$$

where $\theta(\eta)$ is the unit step function, r_{ij} is the distance between two particles i and j ($\underset{\sim}{r}_{ij}=\underset{\sim}{r}_j-\underset{\sim}{r}_i$) and $\phi'(r_{ij})$ is the first derivative of the interaction pair potential with respect to r. Equation (4.15) applies to a surface of any shape. In a planar interface, it is easy to see from the symmetry of the system that the pressure tensor has only two components, $P_N(z)$ and $P_T(z)$, the normal and tangential components, respectively. Hydrostatic equilibrium imposes the conditions that $P_N(z)$ is a constant everywhere, even in the surface, and is equal to the hydrostatic pressure. The tangential component differs from the normal component in the surface region, and it is this component against which work must be done when the surface area of the liquid is changed.

In a computer simulation, these two components of the pressure tensor can be measured [4.32]. For the planar interface, the pressure tensor can be written as

$$\underset{\approx}{P}(\underset{\sim}{r}) = P_T(z)(\underset{\sim}{e}_x\underset{\sim}{e}_x + \underset{\sim}{e}_y\underset{\sim}{e}_y) + P_N(z)\underset{\sim}{e}_z\underset{\sim}{e}_z \quad , \qquad (4.16)$$

where e_x, e_y, and e_z are the unit vectors along the three mutually perpendicular directions. Using (4.15,16), the two components of the pressure tensor in a planar interface can be written as

$$P_N(z) = n(z)k_BT - \frac{1}{2A}\langle\sum_{i\neq j}\frac{|z_{ij}|\phi'(r_{ij})}{|\underset{\sim}{r}_{ij}|}\theta\left(\frac{z-z_i}{z_{ij}}\right)\theta\left(\frac{z_j-z}{z_{ij}}\right)\rangle \qquad (4.17)$$

and

$$P_T(z) = n(z)k_BT - \frac{1}{4A}\langle\sum_{i\neq j}\frac{(x_{ij}^2+y_{ij}^2)}{|r_{ij}||z_{ij}|}\phi'(r_{ij})\theta\left(\frac{z-z_i}{z_{ij}}\right)\theta\left(\frac{z_j-z}{z_{ij}}\right)\rangle \quad , \qquad (4.18)$$

where A is the cross-sectional area of the surface. It is a straightforward procedure to calculate the two components of the pressure tensor from the MD trajectories

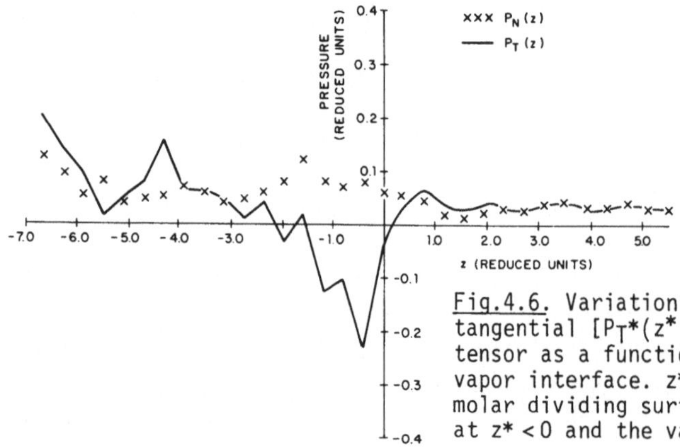

Fig.4.6. Variation of the normal [P_N*(z*)] and tangential [P_T*(z*)] components of the pressure tensor as a function of heigth z* in a liquid-vapor interface. z* = 0 denotes the Gibbs equimolar dividing surface with the liquid phase at z* < 0 and the vapor phase at z* > 0

as ensemble averages [4.32]. Figure 4.6 shows the two components as functions of z obtained from a Monte Carlo simulation of Lennard-Jones liquid-vapor film. The system consists of 2048 particles interacting via a truncated Lennard-Jones (6,12) potential at 100 K. The origin is chosen to be the Gibbs equimolar dividing surface [4.31] given by

$$z_g = \frac{L_z(\bar{n}-\bar{n}_g)}{2(n-n_g)} \quad ,$$

where $\bar{n} = N/V$ is average density of the system, n is the liquid density, and n_g is vapor density. Since the components of the pressure tensor are sensitive to the two-particle distribution function, very lengthy simulations are needed to obtain reasonable statistics.

In the next two sections, we shall discuss how the surface tension and its dependence on the radius of curvature of the interface can be studied using the pressure tensor.

Surface Tension. Statistical mechanics relates the surface tension to the pressure tensor [4.30]. Following reference [4.31], the surface tension γ is given by

$$\gamma = \int_{-L/2}^{L/2} \left[P_N(z) - P_T(z)\right] dz \quad . \tag{4.19}$$

KIRKWOOD and BUFF [4.33], using the above relation, have derived an approximate relation for the surface tension in terms of the pair correlation function of the bulk liquid g(r):

$$\gamma = \frac{\pi n^2}{8} \int_{\infty}^{\infty} \phi'(r) g(r) r^4 dr \quad . \tag{4.20}$$

This relation has been widely used to determine the surface tension of various liquids from the knowledge of the radial distribution function [4.34]. LEE et al. [4.20], on the other hand, obtained the surface tension directly from the computer simulation using (4.19), which can be written as

$$\gamma = \frac{1}{L_x L_y} \langle \sum_{i>j} \sum \frac{(x_{ij}^2 - z_{ij}^2)}{r_{ij}} \phi'(r_{ij}) \rangle \quad . \tag{4.21}$$

Using the same method, CHAPELLA et al. [4.35] and RAO and LEVESQUE also obtained the surface tension of a Lennard-Jones (6,12) fluid. The outcome of these results is that the surface tension is quite sensitive to the long-range part of the interaction potential. When a truncated Lennard-Jones potential is used to reduce the amount of computation involved in the calculation of the potential (or force), the surface tension depends very much on the truncation parameter. Thus, to obtain the correct surface tension with infinite cut-off, "tail corrections" have to be

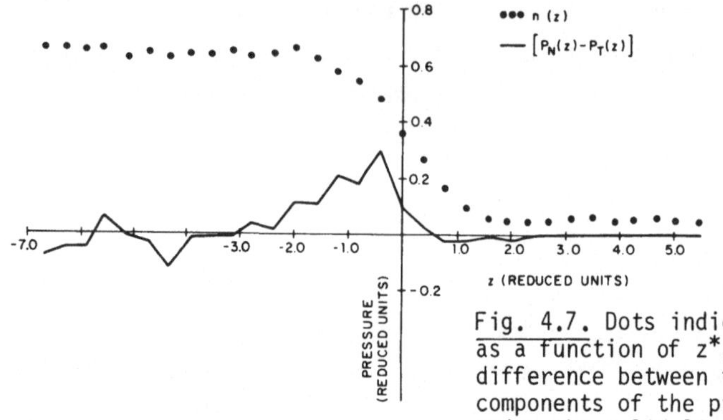

Fig. 4.7. Dots indicate density profile $n^*(z^*)$ as a function of z^*. The solid line shows the difference between the normal and tangential components of the pressure tensor. The area under the solid line gives the surface tension

employed. MIYAZAKI et al. [4.36] have used a novel method to calculate the surface tension from measuring the work done to bring two surfaces together. Using this method they have obtained the surface tension of a Lennard-Jones fluid without truncation. Their results show that the surface tension obtained from the computer simulation studies does not agree with the experimental result. This discrepancy has been attributed to the inadequacy of the pairwise Lennard-Jones interaction potential to represent liquid argon interfaces.

Figure 4.7 shows the difference between the normal and the tangential components of the pressure tensor. Also shown in the same figure is the density profile. The figure shows that the transverse pressure is asymmetric around the Gibbs dividing surface. The area under the curve is the surface tension.

Curvature Dependence of Surface Tension. The asymmetry shown in Fig.4.5 can be described in terms of the first moment, given by

$$z_s = \frac{1}{\bar{\gamma}} \int_{-L/2}^{L/2} [P_N(z) - P_T(z)] z dz \quad , \tag{4.22}$$

This defines the position of the "*surface of tension.*" This surface can also be used to divide the liquid and vapor phases instead of the Gibbs equimolar surface. In a planar interface the surface tension does not depend on the choice of the dividing surface, but in a spherical droplet the use of the dividing surface is essential since precise meaning is given to the concepts of area and curvature. The surface of tension defined by (4.22) plays an important role in spherical droplets. The location of z_s measured from the Gibbs dividing surface ($\delta_\infty \equiv z_g - z_s$) is called the curvature dependence of surface tension in spherical droplets (4.30,31,39). The *Tolman* formula [4.37] relates the surface tension of a spherical droplet to the surface tension of a planar sheet,

$$\gamma(r) = \gamma_\infty \left(1 - \frac{2\delta_\infty}{r}\right) \quad . \tag{4.23}$$

From the results in Fig.4.7 $\gamma = 0.42\ \varepsilon/\sigma^2$, reduced units, and $\delta_\infty = 0.96\,\sigma$.

In a spherical droplet, the pressure tensor will have two components, the radial and the tangential components. It is possible to determine these components with the help of (4.15) using computer simulation and obtain both the surface tension and and the surface of tension satisfying the KELVIN [4.37] relation

$$P_{liq} - P_{gas} = \frac{2\gamma(r)}{r} \quad . \tag{4.24}$$

However, the sensitivity of the pressure tensor makes this calculation very lengthy and expensive. In recent times, spherical droplets have been studied by computer simulation with many insights gained into the thermodynamics of microclusters [4.38,39].

4.2.2 Thermodynamics of Microclusters and Nucleation in a Finite System

Homogeneous nucleation in supersaturated gas depends on an interplay between the bulk and surface properties of microclusters. Although a supersaturated vapor has a higher chemical potential than a bulk liquid of density n_1 and is thereby chemically unstable, this vapor can exist indefinitely in a metastable state. When the supersaturation exceeds a certain value, the vapor condenses sponteneously. In classical nucleation theory [4.38], the Gibbs free energy of formation ΔG_F of a spherical droplet of radius r from an infinite supersaturated gas maintained at a pressure P is given by

$$\Delta G_F = 4\pi r^2 \gamma - \frac{4\pi r^3}{3} n_L k_B T \ln S \quad , \tag{4.25}$$

where $S = P/P(T) > 1$ is the supersaturation and P(T) is the equilibrium vapor pressure at a temperature T. The basic assumptions that lead to equation (4.25) are the following:

1) Macroscopic thermodynamics can be applied to microclusters.
2) The free energy of a microcluster is separable into a bulk and a surface term.
3) The microcluster is spherical.
4) The vapor is an ideal gas.
5) The supersaturation is maintained constant.
6) The surface energy is given by $4\pi r^2 \gamma$.

The fist term in (4.25) represents the work required to create a spherical surface of area $4\pi r^2$, and the second term represents the lowering of the free energy due to the fact that the chemical potential of a bulk liquid is lower than that of a supersaturated vapor. As is well known, the interplay of these surface and bulk terms leads to a free energy maximum, denoted by ΔG_F^*, at a critical radius $r^* =$

$(2\gamma/n_L k_B T \ln S)$. This is the famous barrier to nucleation, and both ΔG_F^* and r^* play a very important role in nucleation theory.

In a computer simulation, however, the total amount of material available for nucleation is fixed (except in a grand canonical ensemble [4.39]). In such a system, the pressure cannot be kept constant during condensation. This requires a modification of the classical nucleation theory and has been the subject of some recent investigations [4.40,41]. These studies show that in a finite system, a "critical droplet size" can still be identified. In addition, there exists a "stable droplet" whose size depends on N, V, and T. If the total free energy of the droplet-vapor system is lower than the free energy of the homogeneous vapor phase, nucleation occurs in a finite system.

Following [4.40], consider a spherical droplet of radius r consisting of $N_1(r)$ "liquid atoms" in equilibrium with $N_g = N - N(r)$ gas atom. The gas atoms are constrained to move in the free volume $V_f = V - V(r)$, where $V = 4\pi(r+\sigma/2)^3/3$ is the excluded volume due to the droplet. The vapor density

$$n_g = \frac{N-N(r)}{V-V(r)} \tag{4.26}$$

and the vapor pressure is given by the equation of state

$$P(r) = n_g k_B T \left[1 + \sum_{k=1}^{\infty} B_{k+1}(T) n_g^k\right] \quad , \tag{4.27}$$

where $B_{k+1}(T)$ is the (k+1)th virial coefficient. The chemical potential of the gas, on the other hand, is given by

$$\mu_g(n_g,T) = \mu_g^0 + k_B T \left[\ln(n_g k_B T) + \sum_{k=1}^{\infty} \frac{(k+1)}{k} B_{k+1}(T) n_g^k\right] \quad , \tag{4.28}$$

and the Gibbs free energy of the gas is

$$G_g = N_g \mu_g(n_g,T) \quad . \tag{4.29}$$

Clearly, if there is no cluster, $n_g \to \bar{n} \equiv N/V$.

Let us assume that the total Gibbs free energy is made up of three parts:

$$G_{total} = G_{gas} + G_{liquid} + G_{surface} \quad , \tag{4.30}$$

where

$$G_{gas} = N_g \mu_g(n_g,T) \quad , \tag{4.31}$$

$$G_{liquid} = N_L \mu_L(n_L,T) \quad , \tag{4.32}$$

$$G_{surface} = 4\pi r^2 \gamma(r) \quad , \tag{4.33}$$

and $\gamma(r)$ is the surface tension of a droplet of radius r.
Note that

$$\mu_L(n_L,T) = \mu_L(n_L^\infty,T) + \frac{1}{n_L^\infty}[P(r)-P_\infty(T)] \quad , \tag{4.34}$$

where $n_L^\infty(T)$ is the density of the macroscopic liquid under its vapor pressure. This follows from an integration of $(d\mu_L)_T = \upsilon_L dp$, assuming that the liquid is incompressible. In equilibrium, we have

$$\mu_L(n_L^\infty,T) = \mu_g(n_g^\infty,T) \quad ; \tag{4.35}$$

hence we obtain

$$G(r) = [N-N_L(r)q]\mu_g(n_g,T) + N_L(r)\mu_g(n_g^\infty) + (n_L^\infty)^{-1}N_L(r)[P(r)-P_\infty(T)] + 4\pi\gamma(r)r^2 \ . \tag{4.36}$$

In a finite system where N, V, and T are constant, the Helmholtz free energy F(r), not the Gibbs free energy, should be minimized to find the conditions for stable equilibrium. The Helmholtz free energy is given by

$$F(r) = G(r) - P(r)V \quad . \tag{4.37}$$

The Helmholtz free energy of formation of the droplet of radius r from an imperfect gas of density $\bar{n}$ at fixed N, V, T is

$$\Delta F_F(r) = G(r) - N\mu_g(\bar{n},T) - [P(r)-P(\bar{n},T)]V \quad , \tag{4.38}$$

where $p(\bar{n},T)$ and $\mu_g(\bar{n},T)$ are the pressure and chemical potential of the initial supersaturated gas of density $\bar{n}$ to be computed from (4.26-28). Figure 4.8 shows the Helmholtz free energy of formation as a function of the droplet size for a Lennard-Jones potential truncated at 2.5. All the parameters needed to compute $\Delta F_F(r)$ are obtained from computer simulation results [4.22,29]. The dots show (4.38) with no

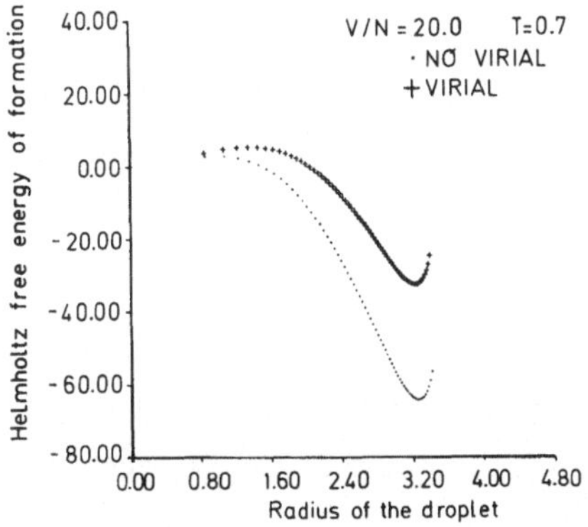

Fig. 4.8. $\Delta F(r)$, the Helmholtz free energy of formation of a droplet of radius r in a Lennard-Jones system for V/N = 20 (in units of σ^3) and a temperature of 84 K. The dots denote the free energy without the virial correction. The crosses denote the free energy with virial correction (2.13). Reduced units are used (energy in units of ε and distances are in units of σ)

virial terms and the crosses show (4.38) with only the second virial correction added. The maxima represent the""critical cluster size," while the minima represent the "stable cluster size." Thus in a finite system, a droplet whose size is given by the minimum will exist in equilibrium with vapor consisting of monomers, dimers, etc. The probability of observing a cluster of radius r is proportional to $\exp[-\beta F(r)]$, so that (4.38) predicts a bimodal distribution of cluster sizes. The stable cluster size itself will depend on the values of N, V, and T decreasing with increasing values of V/N. Eventually for large values of V/N the droplet becomes unstable and only the gas phase will remain. This simple model thus gives an insight into what happened in a computer simulation of a droplet in equilibrium with its own vapor. The theoretical model is somewhat crude and can be refined in a variety of ways.

a) *Preparation of Droplet in Computer Simulation*

Both MD and MC methods have been employed to study the properties of microclusters [4.38-40]. ABRAHAM et al. [4.42] studied clusters of Lennard-Jones fluid surrounded by a spherical region constraining the atoms using the MC method. RAO et al. [4.40], on the other hand, used periodic boundary conditions.

In the studies of RAO et al., the particles are placed in a fully periodic box of size $L \times L \times L$ with a Maxwellian distribution of velocities corresponding to a mean temperature of 0.7 (84 K). This temperature is chosen because the density profile, surface tension, $P_\infty(T)$, etc. in the case of a flat interface are already known from MD simulations for the potential given in (4.12). The integration time step is chosen to be $0.032\,\tau_0$. Starting from a face-centered-cubic lattice configuration, the system is allowed to age to an equilibrium configuration in the periodic box. Next the sides of the periodic cube are increased from L to L', as shown in Fig.4.7, without changing the particle coordinates in any way. This generates an infinite array of cubic droplets (Fig.4.9). The average density $\bar{n}$ thus changes from N/L^3 to N/L'^3. Time evolution is then continued until equilibrium is reestablished. During the initial phases of this time evolution, particles evaporate from the cubical droplets. The temperature of the system decreases during the evolution due to evaporation and the kinetic energy is adjusted periodically (every $\sim$2000 time steps) to maintain a constant temperature of 0.7 at equilibrium. After equilibrium is established (showing constant temperature and pressure within statistical error over 4000 time steps) the evolution is continued further to obtain the density profile n(r) and the energy profile e(r).

The procedure outlined above is also followed using the Monte Carlo method to make sure that what we are observing is indeed an equilibrium state. The usual METROPOLIS et al. [4.12] scheme is used with a step size of 0.2 for the random walk. Within statistical errors, the results obtained by both methods are identical.

However, for large volumes corresponding to small average density, the Monte Carlo method is more efficient since molecular dynamics takes more computer time to stabilize the temperature fluctuations due to evaporation and condensation. On the other hand, molecular dynamics gives the time evolution of the system which is essential in understanding the kinetic processes. Thus both methods are used here to complement each other.

The primary cell in this simulation contained the droplet in equilibrium corresponding to the minimum in the Helmholtz free energy discussed in the previous section. In fact, RAO et al. [4.40] showed that the stable equilibrium can be reached starting from either gas phase and waiting for nucleation to occur, or from a liquid phase with enough free volume so that evaporation will occur. However, to reduce the computation time for equilibration, it is expedient to start with an initial configuration consisting of a liquid droplet and a few gas atoms sprinkled uniformly surrounding it. At equilibrium, the cluster distribution will be bimodal corresponding to the maximum and the minimum in the Helmholtz free energy.

It is important, in order to measure the cluster distribution, to devise a scheme for counting the clusters. A cluster is defined as follows. If any atom lies within a cut-off distance r_c of any other particle, the two particles are said to belong

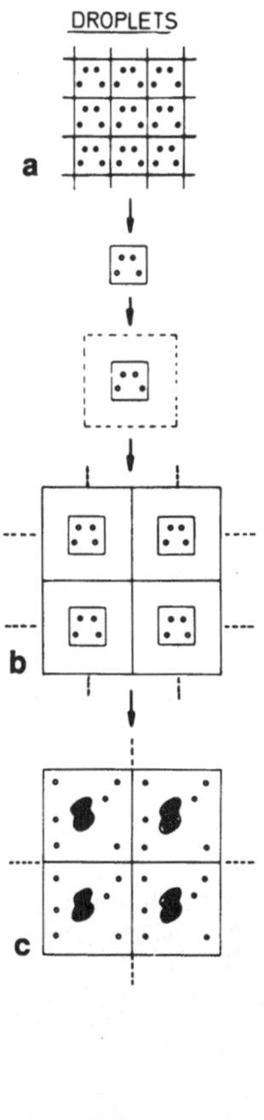

◄Fig. 4.9. (a) Periodic cells of size L containing liquid after equilibrium has been established. (b) Size L is changed to L' without changing the coordinates of the liquid particles, thus creating a vacuum. The evolution is continued until equilibrium is established. (c) Liquid droplet coexisting with vapor after equilibrium is established

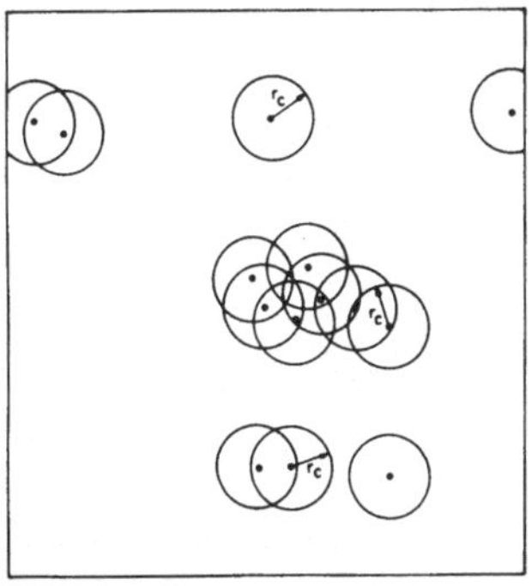

Fig. 4.10. A schematic illustrating the cluster definition in two dimensions

to the same cluster. A schematic illustration of this cluster definition in two dimensions is shown in Fig.4.10. The cut-off distance r_c is adjusted experimentally so that the observed cluster distribution are not too sensitive to the cut-off.

b) *Cluster Distribution*

The average cluster distribution is obtained from an ensemble average. If $N(\ell)$ is the number of clusters of size ℓ, then we have

$$\sum_{\ell} \ell N(\ell) = N \quad . \tag{4.39}$$

The probability of finding a cluster of size ℓ is given by

$$p(\ell) = \frac{N(\ell)}{\sum_{\ell} N(\ell)} \quad . \tag{4.40}$$

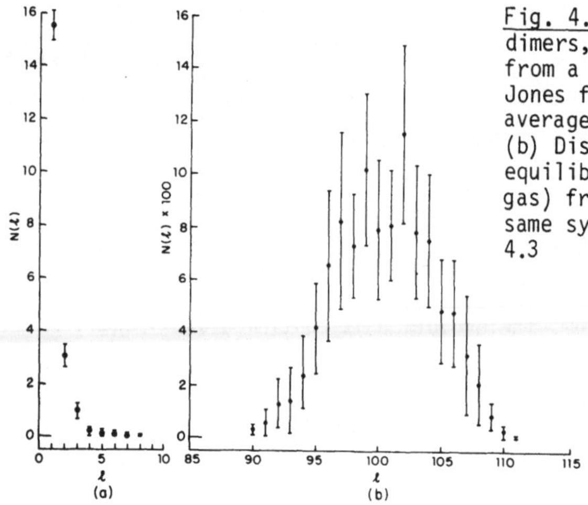

Fig. 4.11. (a) Distribution of monomers, dimers, etc., in the vapor phase obtained from a Monte Carlo simulation of a Lennard-Jones fluid at 84 K corresponding to an average density of $n\sigma^3 = 0.05$ (N = 128). (b) Distribution of large clusters in the equilibrium two-phase system (droplet plus gas) from the computer simulation of the same system as in the dashed lines of Fig. 4.3

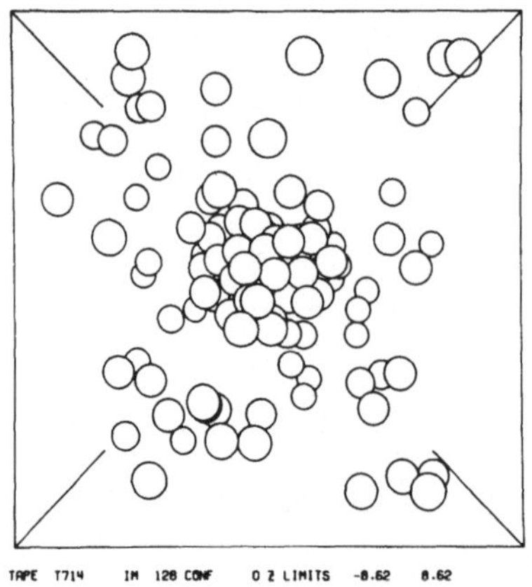

Fig. 4.12. Typical configuration after equilibrium has been established showing stable cluster and vapor. V/N = 40.0, N = 128, T = 0.7

In Fig.4.11 $N(\ell)$ showing the distribution of the clusters as obtained from a simulation of half a million moves after equilibrium has been established (4.40). The theoretical calculation of the Helmholtz free energy presented in Fig.4.8 is based on the same parameters used in this simulation. The stable cluster size predicted with the virial correction included seems to agree well with the experiment. However, certain questions still remain. When $\gamma(r)$ is used instead of γ_∞ for the planar interface, the agreement completely disappears. In addition the droplets are observed to be nonspherical, and the rotation of the droplet must also have an effect on the thermodynamics. These effects have not been considered in the theory. Figure 4.12 shows a frozen snapshot of a typical MD configuration showing the droplet and the vapor.

c) *The Density Profile of a Droplet*

Figure 4.13 shows the density profile of a droplet of 102 particles at 84 K. On the same figure, the density profile of a planar interface of a larger system is shown.

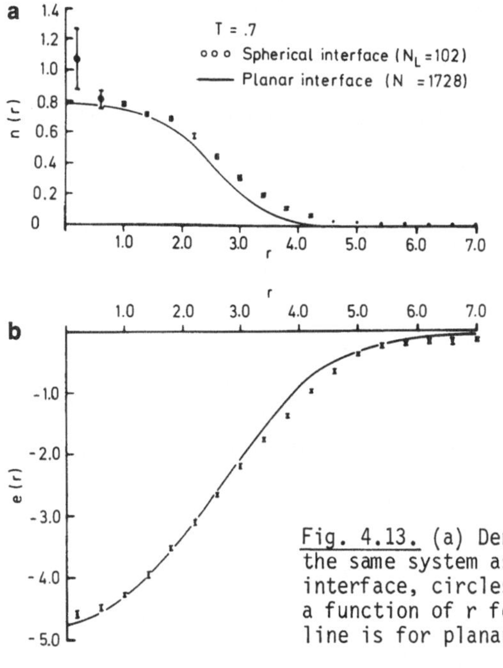

Fig. 4.13. (a) Density profile n(r) as a function of r for the same system as in Fig.4.10a. Solid line is for planar interface, circles are for droplet. (b) Particle energy as a function of r for the same system as in Fig.4.10a. Solid line is for planar interface, circles are for droplet

The widths of the two profiles are strikingly similar, thus suggesting the capillarity picture even in very small size droplets. This picture also is supported by a movie of the droplet in a molecular dynamics simulation. The importance of these capillary waves has only just been recognized and still more work is needed to fully understand their role in nucleation, collision between droplets, and a host of other phenomena.

4.3 Computer Simulation of Idealized Interfaces

Computer simulation of idealized systems have provided an insight into the nature of two-phase boundaries. While the results obtained from these simulations are not directly applicable to real systems, they have several distinct advantages.

a) They are easy to simulate and are "clean" without the complexities present in real systems.

b) They are often amenable to theoretical treatment and thus provide a test for consistency of any theory.

In this section we shall present two examples of such idealized models consisting of a two-phase (liquid-vapor) boundary. These examples are by no means complete, and the reader is referred to literature for further details [4.43].

4.3.1 Lattice Gas Models

This model has been used extensively in providing a wealth of computer simulation data on the thermodynamics and kinetics of a liquid-vapor interface. Furthermore, these results have provided some insight into the mechanism of nucleation, a phenomenon heretofore not accessible to direct study.

An excellent summary of the statistical mechanics of the lattice gas model, and an extensive bibliography are provided in the Appendix of an article by BINDER and KALOS [4.43]. We shall describe a simple two-dimensional lattice gas (square lattice) simulation and give the flavor of the results obtained. The lattice gas is a very simple model of a fluid. The system consists of a square lattice, each lattice site either being occupied or empty. The particles interact only with the nearest neighbors with an interaction energy 4 J. This system has a total energy

$$E = \sum_i \sum_k J\, s_i s_k \quad , \tag{4.41}$$

where $s_i = 1$ if the site is occupied and $s_i = -1$ if the site is empty. The sums run over all the nearest neighbors.

The computer simulation proceeds as follows. A finite number of sites are chosen (e.g., 50 × 50 in two dimensions, or 20 × 20 × 20 in three dimensions, etc.) in a finite size box. Usually periodic boundary conditions are assumed for the box. Within the box, a suitable initial configuration is generated depending on the purpose of the simulation. For example, to study the behavior of critical clusters in a supersaturated vapor [4.43], a fully compact circular cluster with ℓ_i occupied sites is created in the center of the square lattice. Then, N_i gas atoms are chosen to be randomly occupied in the region outside the cluster. This initial condition then defines the average density to be $\bar{n} = (\ell_i+N_i)/L^2$.

The evolution proceeds by choosing at random a pair of nearest-neighbor sites, i and j. If these sites are of opposite kind (one empty and the other occupied), there is a probability that these will be interchanged. This probability is given by (Kawasakian dynamics [4.44])

$$p_{ij} = \frac{\alpha e^{-\beta\delta E_{ij}}}{1+e^{-\beta\delta E_{ij}}} \quad , \tag{4.42}$$

where E_{ij} is the change in the energy of the system caused by the interchange; and $\beta = (kT)^{-1}$.

This rule generates stochastic dynamics in which the total number of particles (occupied sites) is conserved so that $\bar{n}$ is fixed. For temperatures less than the critical temperature T_c in the Ising model, the system evolves in such a way that if the initial gas density $n_G^i = N_i/(L^2-\ell_i)$ is higher than the equilibrium vapor density $n_G^\infty(T)$, gas atoms have to condense on the cluster so that the initial cluster size ℓ_i must be smaller than the equilibrium cluster size ℓ^*. If $n_G^i < n_G^\infty(T)$, atoms must evaporate from the cluster in order to establish equilibrium so that $\ell_i > \ell^*$. Thus when $n_i^G > n_G^\infty(T)$, the system will relax to equilibrium in which a cluster of size ℓ^* is in equilibrium with a vapor consisting of monomers, dimers, etc. It is this equilibrium cluster that is studied and related to the thermodynamic properties of microclusters. A cluster is defined to be an assembly of particles such that no particle is further than one lattice parameter from at least one other particle of the cluster.

This approach to the study of nucleation is very similar if not identical to the study of Lennard-Jones microclusters which precede it by a few years. In fact, even the thermodynamic analysis is very close in spirit. Nevertheless, because the properties of the two-dimensional Ising lattice are known analytically, the lattice gas model allows a more refined determination of the thermodynamic properties, and moreover, by studying the dependence of ℓ^* on $\bar{n}$, the data can be inverted to give the surface free energy and the surface tension as a function of the cluster size ℓ.

4.3.2 Nucleation in Two-Dimensional Square-Well Fluids

The properties of model fluids containing particles interacting with a square-well potential was studied by Alder and Wainwright. Recently, ZUREK and SCHIEVE [4.45-48], in an interesting series of papers, have investigated nucleation in a two-dimensional system of square-well disks. The evolution of the system is followed starting from the simple initial condition where all the particles are placed on a regular lattice with equal speeds but random velocities [4.47-48]. Thus initially ther are no bound molecules. Three body collisions produce dimers, and during the subsequent evolution of the system, larger clusters are formed. The number of dimers, trimers, tetramers, and pentamers were determined as a function of time. Development of a pattern recognition algorithm allowed the authors to study the distribution of different topological structures for each of these oligomers. As in the preceding studies, the system evolved to an equilibrium in which there is a large cluster in equilibrium with dimers, trimers, etc. The authors focused only on these small clusters. They determined the mole fractions of these small cluster using the formula [4.47]

$$m_\ell = \left(\sum_{k_\ell}^{\substack{\text{all clusters}\\ \text{of size}}} t_{k_\ell} \right) \Big/ \left(\sum_{k_0}^{\text{all monomers}} t_{k_0} \right) , \qquad (4.43)$$

where t_{k_ℓ} is the lifetime of the k th cluster of size ℓ, and t_{k_0} is the lifetime of the k th monomer, and found that a plot of $\ell n\, m_\ell$ versus $(kT)^{-1}$ is linear. The activation energy W_ℓ was found to be proportional to the number of bonds in the cluster, but a cluster of given size is often found in noncompact topologies. For example, trimers can be found more frequently in chain rather than triangular configurations. The chains have shorter lifetimes than the triangles, but they occur with higher frequency. This study is provocative and points towards an interesting, more detailed analysis.

ZUREK and SCHIEVE [4.48] called into question the fundamental assumption of nucleation theory, namely, that simple monomer addition

$$A_n + A_1 \rightleftarrows A_{n+1}$$

may not be the major mechanism for growth of the nucleus. Instead, a multistep process

$$A_m + A_n \rightleftarrows A_{n+m}$$

play an important role in condensation and growth of the nucleus. This important result has yet to be tested in three-dimensional systems with more realistic potentials.

4.4 Conclusion

In this chapter we have attempted to give an overview of how various computer simulation techniques have been used to study those statistical properties of heterogeneous systems germane to the study of aerosols, nucleation, etc. This chapter is not meant to be comprehensive. Only a few topics have been included. Nevertheless, it should be clear that computer modeling should be very helpful in the study of heterogeneous systems. Recently, there has been considerable effort expended on the development of a modern theory of heterogeneous systems. PERCUS [4.19] has given an excellent review of progress in this area. The interaction between computer simulations and analytical theory has been very constructive in the development of a theory of uniform fluids. It is to be expected that the same will apply to heterogeneous systems.

In closing, we would like to call the reader's attention to a selection of references relevant to this field but not discussed in this chapter [4.48-101].

Acknowledgements. This work was supported by grants from the National Institutes of Health (9R01 GM 26588-06) and the National Science Foundation (NSF CHE 79-07820).

References

4.1 B.J. Berne, D. Forster: Am. Rev. Phys. Chem. *22*, 563 (1971); W.W. Wood: Acta Phys. Austiaca, Suppl. *X*, 451 (1973)
4.2 B.J. Berne, J. Kushick: In *Statistical Mechanics, Part B: Time-Dependent Processes*, ed. by B.J. Berne (Plenum, New York 1978)
4.3 N. Metropolis, A.W. Metropolis, M.N. Rosenbluth, E. Teller: J. Chem. Phys. *21*, 1087 (1953)
4.4 J.P. Valleau, S.G. Whittington: in *Statistical Mechanics, Part A: Equilibrium Techniques*, ed. by B.J. Berne (Plenum, New York 1977)
4.5 D. Ceperly, J. Tully (ed.), *Stochastic Molecular Dynamics*. Proceedings of Workshop held at Woods Hole, Mass., July 1979. NRCC Proceedings No. 6 and References cited therein.
4.6 S.A. Adelman, J.D. Doll: J. Chem. Phys. *64*, 2375 (1976)
4.7 D. Chandler: J. Chem. Phys. *68*, 2959 (1978); J.A. Montgomery, D. Chandler, B.J. Berne: J. Chem. Phys. *70*, 4056 (1978)
4.8 E. Helfland, Z.R. Wasserman, T.A. Weber: J. Chem. Phys. *70*, 2016 (1979)
4.9 F. Lanteline, P. Turque, H.L. Friedman: In C.E.C.A.M. Workshop on Ionic Liquids, Orsay, France (1974)
4.10 J.P. Hansen, I. MacDonald: *Theory of Simple Liquids* (Academic, New York 1975) [this book contains a detailed discussion of the work by Verlet and co-workers including reference citations]
4.11 D. Forster: *Hydrodynamic Fluctuations, Broken Symmetries and Correlation functions* (Benjamin, New York 1975); A. Rahman, F.H. Stillinger: J. Chem. Phys. *55*, 3336 (1971) F.H. Stillinger, A. Rahman: J. Chem. Phys. *57*, 1281 (1972) A. Rahman, F.H. Stillinger: J. Chem. Phys. *60*, 7943 (1974) F.H. Stillinger, A. Rahman: J. Chem. Phys. *61*, 4973 (1974)
4.12 P. Schofield: Comput. Phys. Comm. *5*, 17 (1973)
4.13 B.J. Berne: in *Statistical Mechanics, Part B: Time-Dependent Processes*, ed. by B.J. Berne (Plenum, New York 1978) p. 233
4.14 J. Tully: J. Chem. Phys. *73*, 1975 (1980) [and references therein]
4.15 C. Pangali, M. Rao, B.J. Berne: Chem. Phys. Lett. *55*, 413 (1978); M. Rao, C. Pangali, B.J. Berne: Mol. Phys. *37*, 1773 (1979)
4.16 J.C. Owicki, H.A. Scheraga: J. Am. Chem. Soc. *99*, 7403, 7413 (1977)
4.17 J.P. Valleau, G.M. Torrie: in *Statistical Mechanics, Part A: Equilibrium Techniques*, ed. by B.J. Berne (Plenum, New York 1977) p. 169
4.18 P. Rossky, J.D. Doll, H.L. Friedman: J. Chem. Phys. *69*, 4628 (1978)

4.19 J.K. Percus: "Non-Uniform Fluids", in *Equilibrium Theory of Liquids*
4.20 J.K. Lee, J.A. Barker, G.M. Pound: J. Chem. Phys. *60*, 1976 (1974)
4.21 K.S. Liu: J. Chem. Phys. *60*, (1976)
4.22 M. Rao, D. Levesque: J. Chem. Phys. *65*, 3233 (1976)
4.23 C.A. Croxton, R.P. Ferrier: Phys. Lett. A *35*, 330 (1971)
4.24 F.H. Abraham, J.K. Lee, J.A. Barker: J. Chem. Phys. *60*, 266 (1974)
4.25 M. Rao, B.J. Berne, J.K. Percus, M.H. Kalos: J. Chem. Phys. *71*, 3802 (1979)
4.26 G.M. Nazarian: J. Chem. Phys. *56*, 1408 (1972);
S. Toxeraera: J. Chem. Phys. *55*, 3116 (1971)
D.G. Sullivan, G. Stell: J. Chem. Phys. *67*, 2567 (1977)
S. Toxeraera: J. Chem. Phys. *64*, 2863 (1976)
4.27 F.F. Abraham: J. Chem. Phys. *68*, 3713 (1978)
H.T. Davis: J. Chem. Phys. *67*, 3636 (1977); *70*, 600 (1979)
4.28 M.S. Wertherm: J. Chem. Phys. *65*, 2377 (1977)
4.29 M.H. Kalos, J. Percus, M. Rao: J. Stat. Phys. *17*, 111 (1977)
4.30 F.P. Buff: J. Chem. Phys. *23*, 419 (1955)
4.31 S. Ono, S. Kondo: in *Structure in Liquids*, Handbuch der Physik, Vol. 10, ed. by S. Flugge (Springer, Berlin, Heidelberg, New York 1960)
4.32 M. Rao, B.J. Berne: Mol. Phys. *37*, 456 (1979)
4.33 J.G. Kirkwood, F.P. Buff: J. Chem. Phys. *17*, 338 (1949)
4.34 R.H. Fowler: Proc. R. Soc. A London *59*, 229 (1937)
4.35 G.A. Chapella, G. Saville, J.S. Rowlinson: Disc. Faraday Soc. *59*, 22 (1975)
4.36 J. Miyazaki, J.A. Barker, G.M. Pound: J. Chem. Phys. *76*, 3364 (1976)
4.37 R.C. Tolman: J. Chem. Phys. *17*, 333 (1949)
4.38 F.F. Abraham: *Homogeneous Nucleation Theory* (Academic, New York 1974)
4.39 J.J. Barton: in *Statistical Mechanics, Part A: Equilibrium Techniques*, ed. by B.J. Berne (Plenum, New York 1977) p. 195
4.40 M. Rao, B.J. Berne, M. Kalos: J. Chem. Phys. *68*, 1325 (1978)
4.41 W. Vogelsberger: Private Communication 1978
4.42 F.F. Abraham, J.K. Lee, J.A. Barker: J. Chem. Phys. *60*, 266 (1974)
4.43 K. Binder, M. Kalos: J. Stat. Phys. *22*, 363 (1980) [and the rather complete list of references cited therein]
4.44 K. Kawasaki: in *Phase Transitions and Critical Phenomena*, Vol. 3, ed. by C. Domb, M.S. Green (Academic, New York 1972)
4.45 W.H. Zurek, W.C. Schieve: Comput. Phys. Commun. *13*, 75 (1977)
4.46 W.H. Zurek, W.C. Schieve: Phys. Lett. *67A*, 42 (1978)
4.47 W.H. Zurek, W.C. Schieve: J. Chem. Phys. *68*, 840 (1978)
4.48 W.H. Zurek, W.C. Schieve: Rarified Gas Dyn. Proc. Int. Symp. *11*, Pt. 2, 1219 (1979)
4.49 J.W. Brady, J.D. Doll, D.L. Thompson: J. Chem. Phys. *74*, 1026 (1981)
4.50 V.V. Nauchitel, A.J. Pertsin: Mol. Phys. *40*, 1341 (1980)
4.51 Mariana Weissman, N.V. Cohan: J. Chem. Phys. *72*, 4562 (1980)
4.52 E. Clementi, G. Coronsiu, B. Joensson, S. Romano: J. Chem. Phys. *72*, 260 (1980)
4.53 M.R. Hoare: Adv. Chem. Phys. *40*, 49 (1979)
4.54 J.S. Turner: Astrophys. Space Sci. *65*, 383 (1979)
4.55 J.Q. Broughton, A. Bonissent, F.F. Abraham: J. Chem. Phys. *74*, 4029 (1981)
4.56 M.R. Mruzik, S.H. Garofalini, G.M. Pound: Surf. Sci. *103*, 353 (1981)
4.57 A. Ueda, J. Takada, Y. Hiwatari: J. Phys. Soc. Jpn. *50*, 307 (1981)
4.58 Y. Yamamura, K. Chikaishi, M. Iaumi, Y. Kitazoe: Kaku Yuso Kenkyu, Bessarau *44*, 33 (1980)
4.59 H.L. Scott, C.Y. Lee: J. Chem. Phys. *73*, 5351 (1980)
4.60 C.Y. Lee, H.L. Scott: J. Chem. Phys. *73*, 4591 (1980)
4.61 C.L. Marquardt, M.E. Ginserich, J.W. Williams: Stud. Surf. Sci. Catal. *4*, 345 (1980)
4.62 K. Binder, D. Stauffer: Adv. Phys. *25*, 343 (1976)
4.63 M.R. Mruzik: Diss. Abstr. Int. B. *36*, 6325 (1976)
4.64 J.B. Kaelberer, Diss. Abstr. Int. B. *36*, 6249 (1976)
4.65 M.J. Mandell, J.P. McTague, A. Rahman: J. Chem. Phys. *64*, 3699 (1976)
4.66 M.R. Mruzik, F.F. Abraham, D.E. Schreiber, G.M. Pound: J. Chem. Phys. *64*, 481 (1976)

4.67 C.L. Briant, J.J. Burton: J. Chem. Phys. *64*, 2888 (1976)
4.68 C.L. Briant, J.J. Burton: J. Chem. Phys. *63*, 2045 (1975)
4.69 K. Binder: J. Chem. Phys. *63*, 2265 (1975)
4.70 F.F. Abraham, J.A. Barker: J. Chem. Phys. *63*, 2255 (1975)
4.71 A. Bonissent, B. Mutaftschiev: J. Cryst. Growth *24*, 503 (1974)
4.72 F.F. Abraham: J. Chem. Phys. *61*, 1221 (1974)
4.73 W. Damgaard Kristensen, E.J. Jensen, R.M.J. Cotterill: J. Chem. Phys. *60*, 4161 (1974)
4.74 J.P. Gayda, H. Ottavi: J. Phys. (Paris) *35*, 393 (1974)
4.75 A.I. Michaels, G.M. Pound, F.F. Abraham: J. Appl. Phys. *45*, 9 (1974)
4.76 D.J. McGinty: J. Chem. Phys. *58*, 4733 (1973)
4.77 J.K. Lee, J.A. Barker, F.F. Abraham: J. Chem. Phys. *58*, 3166 (1973)
4.78 B. Borstnik, A. Azman: Chem. Phys. Lett. *14*, 451 (1972)
4.79 G.H. Gilmer, P. Bennema: J. Cryst. Growth *13*, 144 (1972)
4.80 C.L. Briant, J.J. Burton: J. Chem. Phys. *63*, 3327 (1975)
4.81 C.L. Briant, J.J. Burton: J. Chem. Phys. *63*, 2045 (1975)
4.82 S. Toxvaerd: J. Chem. Phys. *62*, 1589 (1975)
4.83 A.I. Michaels, G.M. Pound, F.F. Abraham: J. Appl. Phys. *45*, 9 (1974)
4.84 H. Meuller-Krumbhaar: in *Monte Carlo Methods*, ed. by K. Binder, Topics in Current Physics, Vol. 7 (Springer, Berlin, Heidelberg, New York 1979) p. 195, 361
4.85 N-H. Tsai: Diss. Abstr. Int. B *39*, 2931 (1978)
4.86 N-H. Tsai, F.F. Abraham, G.M. Pound: Surf. Sci. *77*, 465 (1978)
4.87 J. Miyazaki, G.M. Pound, F.F. Abraham, J.A. Barker: J. Chem. Phys. *67*, 3851 (1977)
4.88 K. Binder: Adv. Colloid Interface Sci. *7*, 279 (1977)
4.89 C. Van Leeuwen, J.P. Van der Eerden: Surf. Sci. *64*, 237 (1977)
4.90 K.W. Mahin, K. Hanson, J.W. Morris, Jr.: Nucl. Metall. *20* (1976)
4.91 S.M. Thompson, K.E. Gubbins: J. Chem. Phys. *70*, 4947 (1979)
4.92 G.H. Gilmer: Faraday Symp. Chem. Soc. *12*, (1977)
4.93 S.M. Thompson, K.E. Gubbins: ACS Symposium Ser. *86*, (1978)
4.94 J.Q. Broughton, L.V. Woodcock: J. Phys. C. *11*, (1978)
4.95 S. Toxvaerd, E. Praestgaard: Ann. Isr. Phys. Soc. *2*, 495 (1978)
4.96 A.I. Rusanov, E.N. Brodskaya: J. Colloid Interface Soc. *62*, 542 (1977)
4.97 G. Saville: J. Chem. Soc. Faraday Trans. *73*, 1122 (1977)
4.98 J.M. Haile: Diss. Abstr. Int. B. *38*, 754 (1977)
4.99 E.N. Brodskaya, A.I. Rusanov: Kolloidn. Zh. *39*, 646 (1977)
4.100 S.P. Protsenko, V.P. Skripov: Fiz.Nizk.Temp. (Kiev) *3*, 5 (1977)
4.101 G.H. Bishop, G.A. Bruggman, R.J. Harrison, J.A. Cox, Sidney Yip: Nucl. Metall. *76* (20)

5. Aerosol Growth by Condensation

P. E. Wagner

With 12 Figures

Formation and evolution in time of natural and anthropogenic aerosols are influenced by gas-to-liquid phase transitions. The formation process of a liquid phase from the vapor can usually be divided into three steps. First, a small amount of the new phase is formed spontaneously (*nucleation*). Secondly, an increasing amount of the new phase accumulates around the initially formed nuclei (*condensational growth*). Further particle growth is finally caused by collision and coalescence of the droplets (*coagulation*). Depending on the actual physical situation, two or even all three of these processes can occur simultaneously.

In the present chapter the condensational growth of aerosol particles is considered. A first-order theory of droplet growth is developed (Sects.5.2-4) and thereby the usually applied theoretical model [5.1-3] is extended. The droplet growth calculations are based on the full, first-order phenomenological equations for heat and mass transport. Thus the influence of the interactions of heat and mass flux on the droplet growth process is determined quantitatively. Different expressions are used to take transitional effects into account. Thermal and mass accomodation at the droplet surface are discussed. In addition the mutual influence of the growing droplets is described quantitatively. Theoretical growth curves, obtained by means of numerical methods, are presented.

Furthermore the results of various experimental investigations are reviewed (Sect.5.5). Particular attention is given to the recent progress achieved by utilizing new light scattering techniques for quantitative observation of fast growth processes. Measurements are described that were performed by means of a process-controlled, fast-expansion chamber, recently developed in the Institute for Experimental Physics of the University of Vienna. A comparison of the numerically obtained droplet growth curves with experimental data is performed (Sect.5.6). Thereby some additional information about the physical processes occuring during the condensational growth of liquid aerosol particles is obtained.

5.1 Statement of the Problem

A system consisting of essentially two different chemical components is considered in which no chemical reactions take place. One component is the carrier gas, which is assumed to be an ideal gas. The other component, the condensate, occurs in both the gaseous (vapor) and the liquid state (droplets). The droplets may be dilute solutions of additional chemical components. The vapor concentration is sufficiently small compared to the concentration of the carrier gas that the vapor can be considered approximately as an ideal gas. The total pressure in the system is constant and uniform. The droplets are spherical and randomly located. For the present investigations, the droplet aerosol is assumed to be monodispersed. However, a generalization to polydispersed systems can be performed. A typical system dimension is assumed to be large compared to the droplet radius. The droplet concentration is sufficiently small, so that the average distance of two neighboring droplets is large compared to the droplet radius. The influence of external forces on vapor diffusion and heat conduction is assumed to be negligible. The fluxes of vapor and heat across the system borders are assumed to be very small.

The validity of the above assumptions depends on the actual physical situation and will be discussed in Sects.5.5,6.

Furthermore, the movement of the droplets relative to the carrier gas is assumed to be sufficiently slow, so that the influence of the convective mass and heat transfer to the droplets can be disregarded. FUCHS [5.4] considered small droplet velocities, where the mass transport due to convection is small compared to the diffusive transport. In this case it is found that an increase of the vapor flux on one side of the droplet is compensated by a decrease of the vapor flux on the other side. Accordingly, at low Reynolds numbers no influence of the droplet movement on the growth rate is expected. From theoretical considerations FRÖSSLING [5.5] obtained the general form of a correction factor for the mass flux to a ventilated drop. FRÖSSLING performed experiments with suspended drops in a wind channel and showed the validity of this correction factor in a range of Reynolds numbers between 2 and 800. These results were applied by SQUIRES [5.6] to calculate the influence of ventilation on the growth rate of cloud droplets falling with terminal velocity. The numerical results of SQUIRES show that for freely falling cloud droplets with a radius less than 10 μm, the deviations from the theory for droplets at rest are smaller than 1% of the droplet radius. The above-mentioned calculations are restricted to droplets in the continuum regime. BROCK [5.7] considered the evaporation of a moving droplet in the transition and free molecule regimes. From the above-mentioned results it can be concluded that the effect of ventilation due to gravitational settling of the droplets is negligible if the droplet radius is assumed to be less than 10 μm.

Finally, the droplet growth or evaporation process is assumed to be quasistationary. Accordingly, it is assumed that the vapor concentration and temperature profiles always correspond to steady-state solutions, and changes during the growth process are determined only by the changing boundary conditions. The range of applicability of the quasi-steady-state assumption has been investigated by a number of authors.

Most studies of nonstationary growth and evaporation are restricted to the continuum regime. LUCHAK and LANGSTROTH [5.8] considered the evaporation of a droplet, situated in the center of a spherical enclosure, whose surface maintains zero vapor concentration. The nonstationary diffusion equation was solved by a method of successive approximation. It was found that the nonstationary correction for the mass flux is less than 1% for water at 20°C. However, the radius of the enclosure must not exceed the droplet radius by more than a factor of about 1700. The heat flux to the droplet was neglected. FRISCH and COLLINS [5.9] solved the nonstationary diffusion equation for a single droplet located in an infinite region with initially constant vapor concentration. After an initial growth period, the well-known quadratic growth law is obtained. A size-dependent accomodation coefficient is considered. The effect of the moving boundary on the droplet growth process was taken into account by KIRKALDY [5.10] by introducing a convection term in the phenomenological equation for the mass flux, whose time dependence was chosen as a trial. The resulting vapor concentration field agrees with the quasi-steady-state solution. However, KIRKALDY concludes that this agreement is caused by a "lucky combination of compounded errors."

CARSTENS and ZUNG [5.11] solved the nonstationary equations for simultaneous mass and heat transport in the continuum regime. The growing droplet is surrounded by an impermeable sphere. The nonstationary solutions show slightly higher gradients of vapor concentration and temperature than in the case of quasi-steady state. However, after an initial growth period of about 10^{-4} s the quasi-steady-state solution is reached. CHANG and DAVIS [5.12] solved the nonstationary continuum equations taking into account the nonuniform temperature field inside the droplet. It is found that droplet evaporation can be approximated by quasi-steady-state equations except for a short initial period. NIX and FUKUTA [5.13] used time-dependent source functions in the nonsteady state continuum equations for mass and heat transfer. Accordingly, changes of the environmental conditions during the growth process, as occuring for example during the expansion period in an expansion chamber, can be taken into account. It is found that the non-steady-state droplet growth "overshoots" the quasi-steady-state conditions during the initial stages of condensation. The relative error R_m for the growth rate caused by the use of quasi-steady-state equations is proportional to the droplet radius. Furthermore R_m decreases with an increasing time constant for the environmental changes (e.g., expansion time, see Sects.5.5,6) and with increasing growth time. For most practically important conditions, a maximum error of about 1% can be expected.

A condition for the applicability of quasi-steady-state solutions has been given by TWOMEY [Ref.5.14,p.71]. TWOMEY assumes that a steady-state vapor concentration field has developed around a droplet. At a particular time the droplet radius is abruptly increased by a certain amount ξ. Calculation of the corresponding non-steady-state flux shows that quasi-steady state is a good approximation if ξ is sufficiently small. This translates into the condition that the vapor density must be much smaller than the liquid density, which is fulfilled in most practical cases. The above-mentioned non-steady-state calculations were based on continuum equations. BROCK [5.15] investigated the case of non-continuum non-steady-state transfer. His results indicate that "a quasistationary assumption is valid in general for aerosol collisions and molecular transfer." Although most theoretical treatments show that the quasistationary theory of droplet growth provides a good approximation, the actual validity of the quasi-steady-state equations can only be checked by comparison with experimental data.

As a consequence of the assumed quasistationarity, the temperature profiles inside the droplets are flat and no heat fluxes into the droplets occur. Accordingly, the influence of the heat capacity of the droplets on the growth or evaporation process is not taken into account.

Without introducing further restrictions, the mass flux to the droplets can only be calculated numerically. Then a numerical integration of the calculated mass flux is necessary in order to obtain the droplet radius as a function of time. A considerable simplification of the calculations can be achieved by assuming that the temperature in the system is nearly uniform, i.e., the temperature profiles are very flat. In this case the temperature of the droplets is nearly equal to the gas temperature at a large distance from the droplets. This condition will be fulfilled approximately for very slow processes and for condensates with small specific latent heat of condensation. With the above restriction it is possible to use a linear expression as an approximation for the vapor pressure versus temperature function. Furthermore a number of parameters, e.g., diffusion coefficient, coefficient of thermal·conductivity, specific latent heat of condensation can be taken as constant and uniform. In this case the mass flux to the droplets can be expressed in analytical form. However, a numerical integration of the mass flux is still necessary to obtain the droplet radius as a function of time.

A further simplification can be obtained by the additional assumption that the droplets are in the continuum regime and that the droplet concentration is sufficiently small, so that the mutual interactions of the droplets can be neglected. In this case the analytical expression for the mass flux can be integrated in closed form and the well-known quadratic growth law is obtained.

5.2 Quasistationary Fluxes to a Single Droplet in the Continuum Regime

Growth and evaporation of liquid droplets are determined by the nonstationary fluxes of vapor and heat to the droplets. These fluxes can only be calculated under certain simplifiying assumptions. In connection with the theory of the wet bulb thermometer, MAXWELL [5.16] was the first to calculate the fluxes of vapor and heat to a single sphere. These fluxes were obtained by solving the stationary equations for diffusion and heat conduction, thereby assuming stationary transport of mass and energy under diffusion control. Accordingly, the expressions obtained are restricted to spheres whose radii are large compared to the mean free path of the surrounding gas (continuum regime). Based on MAXWELL'S results FUCHS [5.4] derived an approximate stationary droplet growth equation taking into account the effect of the latent heat of condensation. MASON [5.2] obtained a somewhat better approximation and took into account the effect of curvature and solution concentration on the vapor pressure at the droplet surface.

For the calculations in this section it will be assumed that the droplet radius is large compared to the mean free path in the surrounding gas (continuum regime). In Sect.5.3 this assumption will be eliminated and appropriate corrections to the continuum expressions will be applied.

5.2.1 Conservation Laws

From the *conservation of mass* in the binary mixture of carrier gas and vapor, the continuity equation

$$\frac{\partial \rho}{\partial t} + \text{div}\, \underset{\sim}{j} = 0 \tag{5.1}$$

can be obtained, where $\rho = \rho_v + \rho_g$ designates the total mass concentration of the binary mixture, $\underset{\sim}{j}$ is the total mass flux density, and t the time. ρ_v and ρ_g are the partial densities of the vapor and the carrier gas, respectively. Because of the assumed quasi-steady state, ρ will not be explicitly time dependent and (5.1) reduces to

$$\text{div}\, \underset{\sim}{j} = 0 \quad . \tag{5.2}$$

The *conservation of energy* in the binary mixture yields the continuity equation

$$\frac{\partial}{\partial t}(\rho_v u_v + \rho_g u_g) + \text{div}\, \underset{\sim}{q} = 0 \quad , \tag{5.3}$$

where u_v, u_g designate the internal energies of vapor and gas per unit mass, respectively, and $\underset{\sim}{q}$ is the heat (energy) flux density. The internal energies can be expressed as

$$u_v = a_v \cdot t + c_{v,v} \cdot T \quad , \tag{5.4}$$

$$u_g = a_g \cdot t + c_{v,g} \cdot T \quad , \tag{5.5}$$

where a_v, a_g are specific chemical reaction rates, $c_{v,v}$, $c_{v,g}$ are specific heat capacities at constant volume, and T is the absolute temperature. Inserting (5.4,5) into (5.3) and assuming that ρ_v, ρ_g, a_v, a_g are constant, the equation

$$(\rho_v \cdot c_{v,v} + \rho_g \cdot c_{v,g}) \frac{\partial T}{\partial t} + \text{div}\, \underset{\sim}{q} = -\rho_v a_v - \rho_g a_g \tag{5.6}$$

is obtained. It can be seen that the chemical reactions cause a source (sink) term in the continuity equation. In the present calculations no chemical reactions are considered. Accordingly, (5.6) reduces to

$$(\rho_v \cdot c_{v,v} + \rho_g \cdot c_{v,g}) \frac{\partial T}{\partial t} + \mathrm{div}\, \underline{q} = 0 \quad . \tag{5.7}$$

For the assumed quasi-steady state, (5.7) further reduces to

$$\mathrm{div}\, \underline{q} = 0 \quad . \tag{5.8}$$

5.2.2 Phenomenological Equations

The conservation laws are not sufficient for a determination of density and temperature profiles. In addition, relations between mass concentration and mass flux density, on the one hand, and between temperature and heat flux density, on the other hand, are required. It has been found that these relations are linear to a high degree of approximation. Accordingly, linear phenomenological expressions are chosen with empirical coefficients. Fortunately, in the case of dilute gases these phenomenological equations can be rigorously derived by solution of the Boltzmann transport equation. For the present calculations, the first-order phenomenological equations obtained by means of ENSKOG's perturbation technique [5.17,18] are used. It should be mentioned that the first-order phenomenological equation for mass transport has been derived by FÜRTH [5.19] using an elementary theory.

By means of the first-order phenomenological equations the mutual interactions of mass and heat flux (e.g., thermal diffusion) can be taken into account, whereas usually mass and heat flux to the droplets are calculated independently in the theory of droplet growth. Furthermore the usual restriction to very low vapor concentrations can be eliminated.

STEFAN [5.20,21] was the first to present a phenomenological equation for mass transport which is somewhat more rigorous than Fick's law. For the validity of STEFAN's equation a restriction to small vapor concentrations is not required. However, STEFAN's equation only applies in the absence of thermal gradients. The solutions, obtained by means of STEFAN's equation, include a description of the additional mass flux from an evaporating surface caused by the center-of-mass motion of the air-vapor mixture relative to the surface. The effect of this Stefan flow on droplet growth and evaporation is small if the vapor concentration is small compared to the gas concentration. VOLKOV and GOLOVIN [5.22] give a description of non-steady-state effects for droplet evaporation, based on the full, first-order phenomenological equations.

The first-order *phenomenological equation for the mass transport* in a binary mixture in the absence of external forces and pressure gradients is usually given in the form

$$\underline{v}_v - \underline{v}_g = -\frac{D}{X_v X_g}(\text{grad } X_v + \frac{k_T}{T}\text{ grad } T) \quad , \tag{5.9}$$

where $\underline{v}_v$, $\underline{v}_g$ are the average diffusion velocities of vapor and gas, respectively. $X_v = N_v/N$ and $X_g = N_g/N$ are the mole fractions of vapor and gas, respectively. $N = N_v + N_g = \rho/M$ is the total molar concentration of the binary mixture. $N_v = \rho_v/M_v$ and $N_g = \rho_g/M_g$ are the molar concentrations of vapor and gas, respectively. M_v, M_g, and M are the molecular weights of vapor, gas, and the binary mixture, respectively. Because vapor and gas are assumed to be ideal gases, the mole fractions can be expressed as $X_v = p_v/p_o$, and $X_g = p_g/p_o$, where $p_o = p_v + p_g$ is the total pressure of the binary mixture and p_v, p_g are the partial pressures of vapor and gas, respectively.

From (5.9) it can be seen that an exchange of vapor and gas does not affect the binary diffusion coefficient D, while the thermal diffusion ratio k_T will change sign. If in (5.9) $k_T > 0$, the diffusion of vapor towards the cooler region and the diffusion of gas towards the warmer region will be enhanced. The dependence of k_T on the mole fractions is approximately given by

$$k_T = \alpha X_v X_g \quad , \tag{5.10}$$

where the absolute value of the thermal diffusion factor α is usually less than 0.6.

The binary diffusion coefficient D is approximately proportional to T^{μ}/p_o, where $1.6 < \mu < 2$ in most cases. The first-order expression for D, as obtained from the rigorous kinetic theory of gases, is independent of the composition of the binary mixture [5.17]. Only higher order corrections are dependent on the composition. Accordingly, because of the assumed small vapor concentration, D can be considered as independent of the mole fractions of vapor and gas.

Because of the assumptions that vapor and gas are ideal and that the gradient of the total pressure can be neglected, (5.9) can be transformed into

$$\underline{v}_v - \underline{v}_g = -\frac{\rho}{\rho_v \rho_g} D^{(v)} \left[\text{grad } \rho_v + \frac{\rho_v + \rho k_T^{(v)}}{T} \text{ grad } T\right] \quad , \tag{5.11}$$

where a modified binary diffusion coefficient $D^{(v)}$ and a modified thermal diffusion ratio $k_T^{(v)}$ are defined as

$$D^{(v)} \equiv \frac{1}{\rho}(\rho_g + \frac{M_g}{M_v}\rho_v) \cdot D = \frac{M_g}{M} \cdot D \quad , \tag{5.12}$$

$$k_T^{(v)} \equiv \frac{1}{\rho}(\rho_v + \frac{M_v}{M_g}\rho_g) \cdot k_T = \frac{M_v}{M} \cdot k_T \quad . \tag{5.13}$$

Because of the assumed low vapor concentration, the modified coefficients $D^{(v)}$, $k_T^{(v)}$ can be approximately expressed as

$$D^{(v)} \cong D \quad , \tag{5.14}$$

$$k_T^{(v)} \cong \frac{M_v}{M_g} \cdot k_T \quad . \tag{5.15}$$

By introduction of the mass average velocity $\underline{v} = (\rho_v \underline{v}_v + \rho_g \underline{v}_g)/\rho$, (5.11) can be transformed into the first-order phenomenological equation

$$\underline{j}_v = -D^{(v)} \left(\text{grad } \rho_v + \frac{\rho_v + \rho k_T^{(v)}}{T} \text{grad } T\right) + \rho_v \cdot \underline{v} \tag{5.16}$$

for the mass flux density $\underline{j}_v = \rho_v \underline{v}_v$ of the vapor. The first term of (5.16),

$$-D^{(v)} \text{ grad } \rho_v \quad ,$$

describes the mass flux of the vapor due to the vapor density gradient. The second term of (5.16),

$$-D^{(v)} \cdot \frac{\rho_v}{T} \text{grad } T \quad ,$$

is caused by the fact that according to (5.9), the driving force for the mass flux is the gradient of the mole fraction X_v, not the gradient of the partial density ρ_v. For the case of a binary mixture of ideal gases at uniform total pressure p_0, it can be seen easily that grad X_v is proportional to grad p_v but not proportional to grad ρ_v. Grad X_v will be proportional to grad ρ_v only in the case of uniform temperature. In this case the above-mentioned term vanishes. The third term of (5.16),

$$-D^{(v)} \cdot \frac{\rho \cdot k_T^{(v)}}{T} \text{grad } T \quad ,$$

describes the effect of thermal diffusion (SORET effect). The fourth term of (5.16),

$$\rho_v \cdot \underline{v} \quad ,$$

describes the mass flux of the vapor caused by the mass average (convective) velocity $\underline{v}$ relative to the observer. This flux is often called Stefan flow.

For the case where temperature gradients and the mass average velocity are negligible, (5.16) reduces to the well-known zeroth-order phenomenological equation

$$\underline{j}_v = -D^{(v)} \text{ grad } \rho_v \quad , \tag{5.17}$$

which is known as Fick's law of diffusion. In this case the above-mentioned first-order effects can be disregarded.

The first-order *phenomenological equation for the heat (energy) transport* in a binary mixture is given in the form

$$\underline{q} = -K \text{ grad } T + \frac{\rho RT}{M} k_T (\underline{v}_v - \underline{v}_g) + \rho_v h_v \underline{v}_v + \rho_g h_g \underline{v}_g \quad , \tag{5.18}$$

where h_v and h_g are the specific enthalpies of vapor and gas, respectively, and R is the universal gas constant.

The coefficient of thermal conductivity K of the binary mixture of vapor and gas depends on the temperature and the mole fractions [5.23]. Based on kinetic theory, WASSILJEWA [5.24] proposed an approximate expression of the form

$$K(T,X_v/X_g) = \frac{K_v(T)}{1+A_{vg}\, X_g/X_v} + \frac{K_g(T)}{1+A_{gv}\, X_v/X_g} \quad . \tag{5.19}$$

The temperature dependence of the thermal conductivities K_v, K_g of vapor and gas can be approximated for a limited temperature range by linear expressions with empirical coefficients. The parameters A_{vg} and A_{gv} are nearly independent from temperature and can be obtained form the semiempirical theory of LINDSAY and BROMLEY [5.25].

The specific enthalpies h_v and h_g can be calculated by means of the expressions

$$h_v(T) = \int_0^T c_{p,v}\, dT' \quad , \tag{5.20}$$

$$h_g(T) = \int_0^T c_{p,g}\, dT' \quad , \tag{5.21}$$

where $c_{p,v}$ and $c_{p,g}$ are the specific heat capacities at constant pressure of vapor and gas, respectively.

The first term of (5.18),

$$-K \text{ grad } T \quad ,$$

describes the heat flux due to the temperature gradient. The second term of (5.18),

$$\frac{\rho RT}{M} k_T \,(\underline{v}_v - \underline{v}_g) \quad ,$$

describes the diffusion thermal effect (Dufour effect). This effect is reciprocal to the effect of thermal diffusion (Soret effect). As a consequence of Onsager's symmetry theorem, the Dufour effect and the reciprocal Soret effect are determined by the same coefficient k_T. The third term of (5.18),

$$\rho_v h_v \underline{v}_v + \rho_g h_g \underline{v}_g \quad ,$$

describes the heat flux, caused by the energy, which is carried with the diffusing vapor and gas molecules.

For the case, where the average diffusion velocities are negligible, (5.18) reduces to the well-known zeroth-order phenomenological equation

$$\underline{q} = -K \text{ grad } T \quad , \tag{5.22}$$

which is known as Fourier's law of heat conduction. In this case the above-mentioned first-order effects can be disregarded.

Fick's law (5.17) and Fourier's law (5.22) are usually applied in the theory of droplet growth. However, Fourier's law of heat conduction only applies in the absence of diffusive mass transport. Furthermore, Fick's law of diffusion is valid only in isothermal gas mixtures, whereas considerable temperature gradients will occur in the neighborhood of growing droplets. Accordingly, it can be expected that Fick's and Fourier's laws will not yield an accurate theoretical description of droplet growth, particularly for liquids with high specific latent heat, e.g., water. Based on the first-order phenomenological equation for mass transport, Sedunov [5.3] concludes that the error caused by the use of Fick's law is negligible for growth processes under atmospheric conditions. This point will be discussed in Sects.5.2.3 and 5.6.

5.2.3 Calculation of Heat and Mass Flux

The continuity equations combined with the phenomenological equations can be solved, and vapor concentration and temperature profiles can be obtained that satisfy the boundary conditions at the droplet surface and at infinity. Because of the assumed quasi-steady state, the continuity equations corresponding to the conservation of mass (5.2) and energy (5.8) are of the general form

$$\mathrm{div}\,\underline{f} = 0 \quad , \tag{5.23}$$

where $\underline{f}$ represents the mass or the heat flux density.

As will be shown below, the first-order phenomenological equations for the mass and heat transport can be brought into the general form

$$\underline{f} = -\delta\,\mathrm{grad}\,\psi \quad , \tag{5.24}$$

where ψ represents the mass concentration or the temperature. It will be shown that the coefficient δ is weakly dependent on ψ. This dependence can be expressed approximately as

$$\delta = \delta_0\,(1+\varepsilon\psi) \quad , \tag{5.25}$$

where δ_0 and ε are constants and $(\varepsilon\psi)^2 \ll 1$.

The boundary conditions at the droplet surface and at infinity are

$$\psi\,(r=a) = \psi_a \quad , \tag{5.26}$$

$$\psi\,(r=\infty) = \psi_\infty \quad , \tag{5.27}$$

where r is the distance from the droplet center and a is the droplet radius. The solution of (5.23-27) is approximately given by

$$\psi(r) = \frac{1}{\varepsilon}\left[1-(1-\varepsilon\psi_a)^{a/r}\,(1-\varepsilon\psi_\infty)^{1-a/r}\right] \quad . \tag{5.28}$$

By means of a series expansion of (5.28) the zeroth-order approximation

$$\psi(r) = \psi_\infty + (\psi_a - \psi_\infty)\,\frac{a}{r} \tag{5.29}$$

can be obtained. Equation (5.29) corresponds to the well-known hyperbolic profiles, which are usually considered in droplet growth theory.

From the first-order profile (5.28) the corresponding flux through the droplet surface can be obtained by means of the phenomenological equation (5.24). By inserting (5.25,28) into (5.24), the flux density at the droplet surface can be calculated. Integration over the droplet surface yields the total flux towards the droplet surface

$$F = \frac{4\pi a \delta_0}{\varepsilon}\left[\ln(1-\varepsilon\psi_a) - \ln(1-\varepsilon\psi_\infty)\right] \quad . \tag{5.30}$$

After expanding as a logarithmic series and retaining the first two terms, the first-order total flux towards the droplet surface can be written as

$$F = 4\pi a\,\frac{\delta(\psi=\psi_\infty) + \delta(\psi=\psi_a)}{2}\,(\psi_\infty - \psi_a) \quad . \tag{5.31}$$

For the case, where the dependence of δ on ψ can be neglected, the arithmetic mean in (5.31) can be replaced by δ and the zeroth-order approximation

$$F = 4\pi a\ \delta \cdot (\psi_\infty - \psi_a) \tag{5.32}$$

can be obtained that is usually applied in the theory of droplet growth.

a) *Heat Flux Q_c in the Continuum Regime*

In the phenomenological equation (5.18) for heat transfer, the average diffusion velocities can be calculated by means of the phenomenological equations (5.11,16) for mass transfer. Taking into account that in the present case $\underline{v}_g = 0$, inserting $D^{(v)}$, $k_T^{(v)}$, and k_T according to (5.14,15,10), respectively, and neglecting higher order terms, the expression

$$\underline{q} = -K \text{ grad } T - D\left[\text{grad } \rho_v + \frac{\rho_v}{T}(1+\alpha) \text{ grad } T\right] \cdot \left(\frac{RT}{M_v}\alpha + h_v\right) \tag{5.33}$$

can be obtained, where α is the thermal diffusion factor.

It can be seen that the heat flux density $\underline{q}$ depends on the temperature gradient as well as on the vapor concentration gradient. In order to separate the problem, a relation between the gradients of temperature and vapor concentration is required. This relation can be obtained from the heat balance at the droplet surface

$$\frac{4\pi}{3}\rho_L\, a^3 c_L\,\frac{dT_a}{dt} = Q + L\, I \quad , \tag{5.34}$$

where Q and I are the total fluxes of heat and mass directed towards the droplet, respectively. ρ_L, c_L and L are the density, specific heat capacity and specific heat of condensation of the liquid, respectively. The temperature dependence of L can be approximated for a limited temperature range by a linear expression with empirical coefficients. T_a is the temperature at the droplet surface. Because of the assumed quasi-steady state, T_a is not explicitly time dependent, i.e., $dT_a/dt = 0$. Accordingly, no heat fluxes into the droplet are considered and the influence of the heat capacity of the droplet is not taken into account. In this case (5.34) yields the balance equation

$$Q + L\,I = 0 \quad . \tag{5.35}$$

From this equation a corresponding relation

$$\underline{q} + L\,\underline{j}_v = 0 \tag{5.36}$$

between the flux densities can be derived. By inserting the zeroth-order phenomenological equations (5.17,22), the relation

$$K\,\mathrm{grad}\,T + L\,D\,\mathrm{grad}\,\rho_v = 0 \tag{5.37}$$

between the gradients of T and ρ_v is obtained.

By means of (5.37), the phenomenological equation (5.33) can be transformed into

$$\underline{q} = -KF_T\,\mathrm{grad}\,T \quad , \tag{5.38}$$

where the *thermal correction factor* F_T is given by

$$F_T \equiv 1 - \left[\frac{1}{L} - \frac{D\rho_v}{KT}(1+\alpha)\right]\left(\frac{RT}{M_v}\alpha + h_v\right) \quad . \tag{5.39}$$

The first correction term in F_T,

$$\left[\frac{1}{L} - \frac{D\rho_v}{KT}(1+\alpha)\right]\frac{RT}{M_v}\alpha \quad ,$$

describes the influence of the diffusion thermo effect on the heat transfer. This correction term includes one expression proportional to α^2, which can usually be neglected. The second correction term in F_T,

$$\left[\frac{1}{L} - \frac{D\rho_v}{KT}(1+\alpha)\right]h_v \quad ,$$

corresponds to the heat flux caused by the energy which is carried with the diffusing vapor molecules. It can be seen that this correction term depends on the thermal diffusion factor α. This is explained by the fact that the heat flux, carried by the diffusing vapor molecules, depends on the diffusive flux of the vapor, and that this mass flux is, in turn, influenced by the thermal diffusion effect.

For atmospheric conditions (air-water vapor mixture, T = 20°C, p_0 = 760 Torr, p_v = 15 Torr), the first correction term in F_T amounts to 0.049 α, the second correction term has the numerical value 0.194 - 0.027 α. Unfortunately, reliable experimental values for α are not available. As mentioned above, usually $|\alpha| < 0.6$ is assumed. Based on experimental data from WHALLEY [5.26] and theoretical investigations by MASON and MONCHICK [5.27], KATZ and MIRABEL [5.28] conclude that α is "probably equal to 0.01" for the mixture of water vapor and air.

From the above-mentioned numerical values it can be seen that the diffusion thermal effect influences the heat flux by only a few percent or even less, depending on the choice of α. Furthermore it is partly compensated by the second correction term. However, the heat carried by the diffusing vapor molecules causes a reduction of the heat flux by nearly 20%. In the equations that are usually applied in droplet growth theory, the above-mentioned corrections are not included.

The phenomenological equation (5.38) is of the general form (5.24). Thus according to (5.31), the first-order total continuum heat flux towards the droplet surface can be expressed as

$$Q_c = 2\pi a \, (K_\infty F_{T,\infty} + K_a F_{T,a}) \, (T_\infty - T_a) \quad , \tag{5.40}$$

where K_a, K_∞ are the thermal conductivities and $F_{T,a}$, $F_{T,\infty}$ the thermal correction factors at the droplet surface and at infinity, respectively.

For the determination of the thermal conductivity K_a of the binary mixture and the thermal correction factor $F_{T,a}$ at the droplet surface, the vapor pressure $p_{v,a}$ at the droplet surface must be calculated. For the present calculations it is assumed that the droplet is in the continuum regime. Therefore the droplet surface is in equilibrium with the vapor phase to a high degree of approximation. In this case, the vapor pressure $p_{v,a}$ at the droplet surface can be calculated from the equilibrium vapor pressure $p_s(T_a)$, taking into account the curvature effect (Kelvin's law), and the solution effect (Raoult's law):

$$p_{v,a} = p_s(T_a) \left[\exp \left(\frac{2M_v \sigma_{LV}}{R\rho_L T_a} \cdot \frac{1}{a}\right)\right] \left(1 - \frac{3m_N M_v i}{4\pi\rho_L M_N} \cdot \frac{1}{a^3}\right) \quad . \tag{5.41}$$

Here σ_{LV} denotes the surface tension and ρ_L the density of the liquid. m_N and M_N are the mass and the molecular weight of a soluble condensation nucleus, i is the Van't Hoff factor. The saturation vapor pressure p_s is usually given as a function of temperature by nonlinear empirical equations. The temperature dependence of σ_{LV} and ρ_L can be approximated for a limited temperature range by linear expressions with empirical coefficients.

In addition to the above-mentioned heat flux Q_c, the heat flux Q_{rad} due to radiation has to be taken into account. According to Stefan-Boltzmann's law, the total radiative heat flux directed towards the droplet is given by

$$Q_{rad} = 4\pi a^2 \sigma\,(\varepsilon_\infty T_\infty^4 - \varepsilon_a T_a^4) \quad , \tag{5.42}$$

where σ is a constant, and $\varepsilon_a, \varepsilon_\infty$ are the emissivities of the droplet surface and the environment, respectively. Assuming $\varepsilon_a = \varepsilon_\infty = 1$, and neglecting higher order terms, (5.42) can be simplified and by comparison with (5.40) it can be estimated that

$$\frac{Q_{rad}}{Q_c} \cong \frac{4\sigma T_\infty^3}{K}\,\hat{\alpha} \quad . \tag{5.43}$$

In most practically important cases, $K/4\sigma T_\infty^3 > 5$ mm.

According to the general assumptions in Sect.5.1, the droplet radius a is restricted to $a < 10\mu m$, so that the influence of gravitational settling on the growth process can be neglected. Therefore it can be seen from (5.43) that the heat transport by radiation is negligible for the assumed experimental conditions.

b) *Mass Flux I_c in the Continuum Regime*

In the phenomenological equation (5.11) for mass transfer, $D^{(v)}$ can be replaced by (5.12), and $k_T^{(v)}$ can be replaced by (5.15,10). Taking into account that in the present case $\underline{v}_g = 0$, the mass flux density $\underline{j}_v = \rho_v \underline{v}_v$ of the vapor can be expressed to a high degree of approximation as

$$\underline{j}_v = \frac{1}{1 - \frac{p_v}{p_0}}\, D\left(\text{grad}\ \rho_v + \frac{\rho_v(1+\alpha)}{T}\,\text{grad}\ T\right) \quad . \tag{5.44}$$

Replacing grad T by (5.37) yields

$$\underline{j}_v = -D\, F_M\, \text{grad}\ \rho_v \quad , \tag{5.45}$$

where the *diffusional correction factor* F_M is given by

$$F_M \equiv \frac{1 - \frac{\rho_v L D}{KT}(1+\alpha)}{1 - \frac{p_v}{p_0}} \quad . \tag{5.46}$$

Because of the assumed low vapor pressure, F_M can be approximated by

$$F_M = 1 + \frac{p_v}{p_0} - \frac{\rho_v L D}{KT}(1+\alpha) \quad . \tag{5.47}$$

The first correction term in F_M,

$$\frac{p_v}{p_0} \quad ,$$

describes the influence of the STEFAN flow on the mass transfer.

The second correction term in F_M,

$$\frac{\rho_v LD}{KT} \quad ,$$

is caused by the fact that the driving force for the mass flux is grad X_v, not grad ρ_v. Grad X_v is proportional to grad ρ_v only for uniform temperature. This condition will be approximately fulfilled for low specific latent heat L and high thermal conductivity K. In this case the above-mentioned term is negligible.

The third correction term in F_M,

$$\frac{\rho_v LD}{KT} \hat{\underline{\alpha}} \quad ,$$

describes the effect of thermal diffusion on the mass flux.

For atmospheric conditions (air-water vapor mixture, T = 20°C, p_0 = 760 Torr, p_v = 15 Torr), the numerical values of the above correction terms are 0.020 for the first term, 0.121 for the second term and 0.121 α for the third term. As mentioned earlier, α is probably equal to 0.01.

It can be seen that the Stefan flow is negligible for atmospheric conditions. Furthermore, the thermal diffusion effect influences the mass flux by only a few percent or even less, depending on the choice of α. However, the correction due to the difference between grad X_v and grad ρ_v amounts to about 12%. Although the latter effect is more important than Stefan flow and thermal diffusion, only the influence of the Stefan flow has been usually estimated in droplet growth theory.

The phenomenological equation (5.45) is of the general form (5.24). Thus according to (5.31), the first-order total continuum mass flux towards the droplet surface can be expressed as

$$I_c = 2\pi a(D_\infty F_{M,\infty} + D_a F_{M,a})\,(\rho_{v,\infty} - \rho_{v,a}) \quad , \qquad (5.48)$$

where D_a, D_∞ are the diffusion coefficients and $F_{M,a}$, $F_{M,\infty}$ the diffusional corrections factors at the droplet surface and at infinity, respectively.

Based on the exponential temperature dependence of D, a somewhat better approximation [5.1] can be achieved as follows. Inserting the phenomenological equation (5.45) into the steady-state continuity equation (5.2) and neglecting grad F_M yields the equation

$$\text{grad}\,\rho_v \cdot \text{grad}\,D + D \cdot \Delta\rho_v = 0 \quad . \qquad (5.49)$$

Taking into account the exponential temperature dependence of D, the solution of (5.49) which satisfies the boundary conditions (5.26,27) can be approximately obtained in the form

$$\rho_V = \rho_{V,\infty} + \frac{\rho_{V,a} - \rho_{V,\infty}}{T_a^{1-\mu} - T_\infty^{1-\mu}} \left\{ \left[T_\infty + (T_a - T_\infty) \frac{a}{r} \right]^{1-\mu} - T_\infty^{1-\mu} \right\} \quad . \tag{5.50}$$

Inserting (5.50) into (5.45) yields the flux density at the droplet surface. By integrating over the droplet surface and taking the arithmetic mean of F_M, the expression

$$I_c = 2\pi a\, D_\infty \left[\frac{T_a^{\mu-1}}{T_\infty} \cdot \frac{T_\infty - T_a}{T_\infty^{\mu-1} - T_a^{\mu-1}} \cdot (\mu-1) \right] \cdot (F_{M,\infty} + F_{M,a}) \cdot (\rho_{V,\infty} - \rho_{V,a}) \tag{5.51}$$

for the first-order total continuum mass flux towards the droplet surface can be obtained. If $\mu = 2$, (5.51) reduces to

$$I_c = 2\pi a \sqrt{D_\infty \cdot D_a}\, (F_{M,\infty} + F_{M,a})\, (\rho_{V,\infty} - \rho_{V,a}) \quad . \tag{5.52}$$

5.3 Quasistationary Fluxes to a Single Droplet in the Transition Regime

Up to this point, the calculations were based on the assumption that the droplet is in the continuum regime. In several practically important systems, however, the mean free path λ of the surrounding gas cannot be neglected compared to the droplet radius a. In this case, corrections to the continuum fluxes can be applied that are based on rigorous theory or on semiempirical interpolation techniques.

In the case of a droplet whose radius is much smaller than the mean free path of the surrounding gas (free molecule regime), the diffusion theory predicts mass fluxes at the droplet surface that are much higher than the rate of evaporation. In this case the rate of evaporation is too small to keep the droplet surface saturated and the equilibrium is disturbed. Accordingly, the transport is under kinetic control and the stationary fluxes of mass and energy can be calculated according to kinetic theory.

A difficult situation arises if the droplet radius is of the order of the mean free path of the surrounding gas (transition regime). In this case the transport of mass and energy is partly under diffusion control and partly under kinetic control. At the surface a jump of temperature and vapor concentration occurs, which was pointed out by LANGMUIR [Ref.5.29,p.426]. SCHÄFER [5.30] attempted to calculate the transitional mass flux by equating the stationary continuum and free molecule fluxes at the droplet surface. By matching the fluxes at a jump distance Δ outside the droplet (model of the boundary sphere) FUCHS [5.4] obtained an expression for the stationary mass flux in the transition regime. FUCHS [5.31] argued that Δ should be chosen somewhat larger than the mean free path in the surrounding gas. The actual value of Δ has been estimated by BRADLEY et al. [5.32] and WRIGHT [5.33]. CARSTENS and KASSNER [5.34] and FUKUTA and WALTER [5.35] obtained approximate expressions for the mass and heat flux in the transition regime by equating the continuum and kinetic fluxes at the droplet surface, thereby neglecting FUCHS' jump distance Δ. Based on these results FUKUTA and WALTER [5.35] and CARSTENS et al. [5.36] derived approximate stationary droplet growth equations taking into account the jump of vapor concentration as well as the temperature jump at the droplet surface. Because the jump distance Δ was neglected, the growth equation will be valid only for droplets with radius > 1μm. Recently DAHNEKE [5.37] proposed a simple kinetic theory of the mass and heat transport in the transition regime. Using the continuum theory with certain boundary conditions at the droplet surface (kinetic boundary conditions), DAHNEKE obtains expressions that are in agreement with FUCHS' semiempirical interpolation if the Δ of WRIGHT [5.33] is chosen.

The above-mentioned semiempirical interpolations are limited by the insufficient knowledge of the jump distance and are therefore restricted to sufficiently large droplets. In order to obtain growth equations which are valid for arbitrary droplet sizes, a solution of the rigorous Boltzmann transport equation must be attempted. Because of the complexity of this equation, most authors use a linearization of the collision term suggested by BHATNAGAR et al. [5.38] (BGK model). This approximation was first used by BROCK [5.39] to obtain a first-order correction to the free molecule flux valid in the near free molecule regime. SAHNI [5.40] achieved a solution of the BOLTZMANN equation in order to calculate the neutron flux intensity onto a spherical absorber (black sphere). These results were generalized by SMIRNOV [5.41] for the case of a partially absorbing sphere (grey sphere) and were used in the theory of droplet growth. SMIRNOV obtained expressions for the mass and heat flux to a droplet of arbitrary size and with arbitrary values of the mass and thermal accomodation coefficient. These expressions were used to derive an approximate stationary droplet growth equation for droplets of arbitrary size. SAHNI [5.40] and SMIRNOV [5.41] assumed that the magnitude of the velocity of the molecules is constant. This assumption was eliminated by LOYALKA [5.42]. Another important generalization is due to WILLIAMS [5.43], who took into account anisotropic scattering of the vapor molecules thereby allowing for persistence of velocity. Accordingly, in WILLIAMS' treatment it is not necessary to assume the mass of a vapor molecule to be small compared to the mass of a gas molecule.

A somewhat different approach was chosen by SHANKAR [5.44], who solved the Boltzmann equation by means of the Maxwell moment method. The advantage of this method lies in the fact that the Boltzmann equation need not be solved directly but is solved only for the lower moments of the distribution function. These moments correspond to macroscopic quantities. Numerical calculations show excellent agreement with FUCHS' Δ method if Δ is set equal to zero. The calculations [5.42-44] result in expressions for the mass flux to the droplet, but the heat flux has not been calculated. LOYALKA and WILLIAMS assume an isothermal carrier gas. However, in general, the heat released during the phase transition will cause temperature gradients in the vicinity of the droplet, and explicit expressions for the heat flux to the droplet are required for droplet growth calculations.

5.3.1 Knudsen Numbers

The transitional correction depends on the Knudsen number

$$\mathrm{Kn} \equiv \frac{\lambda}{a} \quad , \tag{5.53}$$

where λ is the mean free path in the surrounding gas and a is the droplet radius. The mean free path in a gas is well defined only in the case where the molecules can be approximated as rigid particles. For a pure gas in the rigid elastic sphere model the mean free path is given [5.18]

$$\lambda = \frac{1}{\sqrt{2}n\pi\sigma^2} \quad , \tag{5.54}$$

where n is the number of molecules per unit volume, σ is the diameter of a molecule, and hence $\pi\sigma^2$ is the collision cross section. In the binary mixture of vapor and gas, the mean free path λ_v of the vapor molecules will be different, in general, from the mean free path λ_g of the gas molecules. It can be shown [5.45] that

$$\lambda_v = \frac{1}{\sqrt{2}n_v\pi\sigma_v^2 + \sqrt{1+m_v/m_g}\,n_g\pi\sigma_{vg}^2} \quad , \tag{5.55}$$

$$\lambda_g = \frac{1}{\sqrt{2}n_g\pi\sigma_g^2 + \sqrt{1+m_g/m_v}\,n_v\pi\sigma_{vg}^2} \quad , \tag{5.56}$$

where n_v, n_g are the numbers of molecules per unit volume, m_v, m_g are the molecular masses, and σ_v, σ_g the molecular diameters of vapor and gas, respectively. $\pi\sigma_{vg}^2 = (\pi/4)(\sigma_v + \sigma_g)^2$ is the collision cross section for unlike-molecule interactions. $\pi\sigma_v^2$ and $\pi\sigma_g^2$ are the collision cross sections for like-molecule interactions for vapor and gas molecules, respectively. Because of the assumed low vapor concentration, (5.55,56) reduce to

$$\lambda_v = \frac{1}{\sqrt{1+m_v/m_g}\,n_g\pi\sigma_{vg}^2} \quad , \tag{5.57}$$

$$\lambda_g = \frac{1}{\sqrt{2}n_g\pi\sigma_g^2} \quad . \tag{5.58}$$

For the experimental determination of the mean free path λ, relations between λ and the transport coefficients can be applied. From simple kinetic theory the zeroth-order approximations [5.17]

$$\lambda = \frac{3D}{\bar{c}} , \tag{5.59}$$

$$\lambda = \frac{3K}{\rho c_v \bar{c}} \tag{5.60}$$

can be obtained for pure gases, where $\bar{c} = \sqrt{8kT/\pi m}$ is the average absolute velocity of the gas molecules. $\rho = nm$ is the mass concentration of the gas, c_v is the specific heat capacity at constant volume, m is the molecular mass, and k is Boltzmann's constant. These approximate equations can be used to determine λ from D or K. Strictly speaking, in (5.59) D is the coefficient of self-diffusion.

In the rigorous theory of transport phenomena, the persistence of velocities has to be taken into account. The first-order kinetic theory formulae for the transport coefficients D, K, and η for diffusion, thermal conductivity and viscosity, respectively, of a pure gas are [5.17]

$$D = \frac{3}{8\Omega^{(1,1)*}} \frac{\sqrt{\pi mkT}}{\pi\sigma^2} \cdot \frac{1}{\rho} , \tag{5.61}$$

$$K = \frac{25}{32\Omega^{(2,2)*}} \frac{\sqrt{\pi mkT}}{\pi\sigma^2} c_v \frac{9\kappa - 5}{10} , \tag{5.62}$$

$$\eta = \frac{5}{16\Omega^{(2,2)*}} \frac{\sqrt{\pi mkT}}{\pi\sigma^2} , \tag{5.63}$$

where $\Omega^{(1,1)*}$ and $\Omega^{(2,2)*}$ are reduced collision integrals that are equal to unity for the rigid sphere model. Numerical values for other intermolecular potentials are tabulated, e.g., [5.17].

$\kappa = c_p/c_v$ is the ratio of specific heat capacities, $(9\kappa-5)/10$ is the semiempirical Eucken correction for polyatomic gases, which is equal to unity for monoatomic gases. The expressions for K and η cannot be simply generalized to binary gas mixtures. However, direct generalization of the formula for D yields the correct first-order expression for the binary diffusion coefficient

$$D_{vg} = \frac{3}{8\Omega_{vg}^{(1,1)*}} \frac{\sqrt{\pi k^3 T^3 (m_v + m_g)/2m_v m_g}}{\pi\sigma_{vg}^2} \frac{1}{p_0} , \tag{5.64}$$

which reduces to (5.61) if $m_v = m_g = m$ and $\sigma_{vg} = \sigma_v = \sigma_g = \sigma$.

According to (5.64) the binary diffusion coefficient is independent of the composition of the binary mixture to the first order of approximation, as already mentioned earlier. It can be seen that the mean free path λ does not appear naturally in the first-order equations (5.61-64).

By means of (5.57) and (5.64), the expression

$$\lambda_v = \frac{32\Omega_{vg}^{(1,1)*}}{3\pi(1+m_v/m_g)} \frac{D_{vg}}{\bar{c}_v} \tag{5.65}$$

can be derived [5.46], where $\bar{c}_v$ is the average absolute velocity of the vapor molecules. Similarly, (5.58) and (5.62) yield the expression

$$\lambda_g = \frac{64\Omega_g^{(2,2)*}}{25\pi} \frac{10}{9\kappa - 5} \frac{K_g}{\rho_g c_{v,g} \bar{c}_g} , \tag{5.66}$$

where $\bar{c}_g$ is the average absolute velocity of the gas molecules. The first-order equations (5.65,66) can be used to determine more accurate values of λ_v and λ_g from the measured quantities D_{vg} and K_g. Comparison of (5.59,60) and (5.65,66) shows that (5.59) is a fairly good approximation for the mean free path of vapor molecules if $m_v \ll m_g$ and $\Omega_{vg}^{(1,1)*} \cong 1$. However, (5.60) is quite different from (5.66). In this connection it is important to note that (5.59,60) were derived only from simple kinetic theory for the case of self-diffusion and heat conduction in a pure gas.

The higher order approximations for D are complicated and depend on the composition of the binary mixture and other parameters. However, for most binary mixtures these corrections do not exceed 3% [5.17]. Expressions for K and η of a pure gas have been calculated up to the fourth order of approximation assuming the rigid elastic sphere model. It turns out that the first-order expressions for K and η according to (5.62,63) should be multiplied by the factors 1.02513 and 1.01600, respectively [5.18]. The fourth order approximations may be taken as accurate to within 0.1%. It can be concluded that the error of the first-order approximations (5.61-64) will not exceed 3%.

In the theory of droplet growth in the transition regime the transitional corrections for the mass and heat flux to a droplet will be dependent on different Knudsen numbers. The transitional correction for the mass flux will be dependent on the Knudsen number

$$Kn_M \equiv \lambda_v/a \tag{5.67}$$

with respect to the vapor molecules. Because of the low vapor concentration it is usually assumed that the transitional correction for the heat flux is dependent on the Knudsen number

$$Kn_T \equiv \lambda_g/a \tag{5.68}$$

with respect to the gas molecules. These Knudsen numbers can be calculated by means of the first-order expressions (5.65,66) for λ_v and λ_g. However, most authors use expressions for Kn_M and Kn_T based on the zeroth-order equations (5.59,60) with slight modifications sometimes applied. For comparison of different transitional corrections, the particular definitions of Kn_M and Kn_T, as chosen by the authors, have to be taken into consideration. The Knudsen numbers based on the first-order expressions (5.65,66) should be used as a common reference [5.46].

For the calculations described here, the corrections due to FUKUTA and WALTER [5.35] and SMIRNOV [5.41] were chosen. The expressions for the Knudsen numbers used in both investigations are direct consequences of the zeroth-order approximations (5.59,60).

5.3.2 Expressions for Mass and Heat Flux

The mass and heat flux in the transition regime can be expressed as

$$I_T = \beta_M I_C \quad , \tag{5.69}$$

$$Q_T = \beta_T Q_C \quad , \tag{5.70}$$

where I_C, Q_C are the continuum fluxes and β_M, β_T are *transitional* correction factors for the continuum fluxes of mass and heat, respectively. In general, β_M and β_T are dependent on the Knudsen numbers Kn_M and Kn_T and on the accomodation coefficients α_M and α_T for mass and energy, respectively. The mass accomodation coefficient α_M is the probability for a vapor molecule hitting the liquid-gas interface to stay within the liquid. The energy (thermal) accomodation coefficient α_T is the probability for a gas molecule hitting the liquid-gas interface to come into thermal equilibrium with the liquid before being diffusely reflected.

According to (5.51,40) the continuum fluxes I_C and Q_C are proportional to D and K, respectively. In order to express the transitional fluxes I_T and Q_T in the same general form, "compensated" transport coefficients $D^* \equiv \beta_M D$ and $K^* \equiv \beta_T K$ are sometimes introduced. Here this notation will not be adopted.

Various expressions for the transitional correction factors have been proposed. FUKUTA and WALTER [5.35] obtained expression for β_M and β_T, which can be written in the form

$$\beta_M^{(F)} = \frac{1}{1+4\ Kn_M/3\alpha_M} \quad , \tag{5.71}$$

$$\beta_T^{(F)} = \frac{1}{1+4\ Kn_T/3\alpha_T} \quad , \tag{5.72}$$

where Kn_M and Kn_T are defined by (5.59,60). It can be expected that (5.71,72) are approximately valid at low Knudsen numbers.

Based upon SAHNI's solution [5.40] SMIRNOV [5.41] obtained the expressions

$$\beta_M^{(s)} = \frac{1}{1+[\Lambda(Kn_M) - 4/3 + 4/3\alpha_M]Kn_M} \quad , \tag{5.73}$$

$$\beta_T^{(s)} = \frac{1}{1+[\Lambda(Kn_T) - 4/3 + 4/3\alpha_T]Kn_T} \quad , \tag{5.74}$$

where Kn_M and Kn_T are defined by (5.59,60) and Λ is a monotonic function of Kn. Numerical values of Λ have been calculated [5.40] and are tabulated for different values of Kn. The limiting values of Λ are given by

$$\Lambda = 0.710 \text{ for } Kn \ll 1 \quad ,$$

$$\Lambda = 4/3 \text{ for } Kn \gg 1 \quad .$$

FUCHS [5.47] proposed the interpolation formula

$$\Lambda(Kn) = \frac{(4/3)Kn + 0.710}{Kn + 1} \tag{5.75}$$

which results in Λ values deviating from SAHNI's values by less than 3%. From (5.73-75) the expressions

$$\beta_M^{(s)} = \frac{1}{(0.377 . Kn_M + 1)/(Kn_M + 1) + 4\ Kn_M/3\alpha_M} \quad , \tag{5.76}$$

$$\beta_T^{(s)} = \frac{1}{(0.377 . Kn_T + 1)/(Kn_T + 1) + 4\ Kn_T/3\alpha_T} \quad , \tag{5.77}$$

can be derived. Equations (5.76,77) can be applied for arbitrary values of Kn. However, strictly speaking, the validity is restricted to isotropic scattering, i.e., $m_v \ll m_g$.

Comparison between (5.71,72) and (5.73,74) shows that the expressions of FUKUTA and WALTER [5.35] and SMIRNOV [5.41] differ by the term $(\Lambda - 4/3)Kn$ in the denominator. As can be seen from above, $\Lambda \leq 4/3$. Therefore the values of the transitional fluxes according to FUKUTA and WALTER will be somewhat smaller than SMIRNOV's values. For large Knudsen numbers, Λ approaches 4/3, but the term $(\Lambda - 4/3)Kn$ does not approach zero.

From the above equations it can be seen that for very small Knudsen numbers, β_M and β_T approach unity and the fluxes become independent from the accomodation coefficients. This is explained by the fact that for the case of negligible mean free path the droplet growth process is under diffusion control and surface kinetic effects have negligible influence. Accordingly, measurements of the accomodation coefficients cannot be performed in the continuum regime.

Because of a modification of the distribution function caused by the density gradient near the droplet surface, the quantities α_M, α_T occuring in the above equations are not identical with the true accomodation coefficients α_M', α_T'. It can be shown that $\alpha_M = \alpha_M'/(1 - \alpha_M'/2)$ with a similar expression for α_T. However, mass and heat flux will be significantly affected by the accomodation coefficients only if α_M, α_T are sufficiently smaller than unity. In this case the above distinction is unimportant. Therefore α_M and α_T are treated here as if they were the actual microscopically significant quantities.

For a droplet with 0.5 µm radius under atmospheric conditions, the Knudsen numbers according to (5.59,60) are

$Kn_M = 0.25$,

$Kn_T = 0.32$.

Under the assumption that the accomodation coefficients α_M and α_T are equal to unity, the numerical values of the transitional correction factors are

$\beta_M^{(F)} = 0.75$, $\beta_T^{(F)} = 0.70$ (FUKUTA, WALTER),

$\beta_M^{(s)} = 0.83$, $\beta_T^{(s)} = 0.78$ (SMIRNOV).

It can be seen that the transitional corrections for a 0.5 μm droplet are of the same order as the diffusional and thermal corrections calculated in Sect.5.2. With increasing droplet radius the transitional correction factors approach unity and thus become less important, whereas the diffusional and thermal corrections are size independent. However, the diffusional and thermal corrections have not yet been included in the usually applied droplet growth theory.

The accomodation coefficients play the role of adjustable parameters in droplet growth theory. For most liquids it has been found that α_M and α_T are close to unity [5.48,49]. However, for a number of liquids, particularly water, values much less than unity have been measured and severe disagreement between different investigations can be observed. In a series of papers starting in 1931, ALTY and coworkers report investigations of the accomodation coefficients of water. Finally, ALTY and MACKAY [5.50] conclude from their measurements that $\alpha_M = 0.036$ and $\alpha_T = 1.0$. HICKMAN [5.51] found α_M to be "not less than" 0.25. In a reanalysis of HICKMAN's data MILLS and SEBAN [5.52] found that $\alpha_M = 1$ "is indeed the best conclusion from the experimental data." Values of α_M between 0.0265 and 0.0415 were found by DELANEY et al. [5.53]. The experiments of MILLS and SEBAN [5.52] indicate that $\alpha_M > 0.45$. AKOY [5.54] finds α_M between 0.55 and 1.17 with a mean value at 0.82. AKOY discussed the possible errors caused by surface contamination, by the presence of a diffusion barrier, and by difficulties involved in measuring the temperature of the surface layer. Measurements by SINNARWALLA et al. [5.55] result in $\alpha_M = 0.022$ to 0.032 and $\alpha_T = 1$. SINNARWALLA et al. [5.55] and NARUSAWA and SPRINGER [5.56] discuss surface contamination as a possible source of error. NARUSAWA and SPRINGER [5.56] obtain $\alpha_M = 0.038$ for a stagnant surface and $\alpha_M = 0.17$ to 0.19 for a replenished (moving) surface.

Based on a comparison of theoretical and experimental droplet growth curves, an estimate for the accomodation coefficients of water is given in Sect.5.6.

5.3.3 Jumps of Density and Temperature

According to (5.69,70), the transitional fluxes I_T and Q_T differ from the continuum fluxes I_c and Q_c by factors β_M and β_T, respectively. However, at distances large

compared to the mean free path in the surrounding gas, the zeroth-order profiles
of vapor concentration and temperature will be similar to the continuum profiles
and can be expressed as

$$\rho_v(r) = \rho_{v,\infty} + \beta_M(\rho_{v,a} - \rho_{v,\infty})\frac{a}{r} \quad , \tag{5.78}$$

$$T(r) = T_\infty + \beta_T(T_a - T_\infty)\frac{a}{r} \quad . \tag{5.79}$$

Inserting the profiles (5.78,79) into the first-order phenomenological equations
(5.45,38) and integrating over the droplet surface yields the expressions

$$I_T = 4\pi a \, D\beta_M F_M(\rho_{v,\infty} - \rho_{v,a}) \quad , \tag{5.80}$$

$$Q_T = 4\pi a \, K\beta_T F_T(T_\infty - T_a) \quad , \tag{5.81}$$

for the total fluxes of mass and heat towards the droplet surface. Equations
(5.80,81) agree with the first-order transitional expressions (5.69,70) in connection with (5.48) and (5.40) if D, F_M, K, and F_T are assumed to be uniform.

The profiles (5.78,79) will only be valid at sufficiently large distances from
the droplet surface. At distances of the order of the mean free path in the surrounding gas, deviations from the hyperbolic profiles (5.78,79) will occur. These deviations can be estimated by extrapolating the profiles (5.78,79) towards the droplet surface and comparing the extrapolated values $\rho_v(a)$, $T(a)$ with the actual values
of vapor concentration $\rho_{v,a}$ and temperature T_a at the droplet surface. It can be
seen that *jumps* of the vapor concentration,

$$\Delta_M \equiv \rho_v(a) - \rho_{v,a} = (\rho_{v,\infty} - \rho_{v,a})(1 - \beta_M) \quad , \tag{5.82}$$

and the temperature,

$$\Delta_T \equiv T(a) - T_a = (T_\infty - T_a)(1 - \beta_T) \quad , \tag{5.83}$$

will occur close to the droplet surface. Equations (5.71-74) show that β_M, β_T decrease with increasing Knudsen numbers. Accordingly, the jumps of vapor concentration Δ_M and temperature Δ_T become more pronounced at higher Knudsen numbers. On
the other hand, as mentioned earlier, β_M and β_T approach unity for very small Knudsen numbers. In this case the jumps Δ_M and Δ_T become insignificant and the hyperbolic continuum profiles extend all the way to the droplet surface.

5.4   Quasistationary Droplet Growth and Evaporation

5.4.1   Mass Flux to a Single Droplet

The growth rate of a droplet can be determined from the total mass flux to the
droplet. For calculation of the total transitional mass flux I_T from (5.69,51),

the temperature T_a at the droplet surface must be inserted. T_a can be obtained from the heat balance at the droplet surface. According to (5.35), for quasi-steady-state conditions the heat flux Q_T and the mass flux I_T are related by the equation

$$Q_T(a,T_a) + L(T_a) \cdot I_T(a,T_a) = 0 \quad , \tag{5.84}$$

where $L(T_a)$ is the specific latent heat of condensation at the droplet temperature. After replacing Q_T by (5.70) and (5.40) and I_T by (5.69) and (5.51), (5.84) can be solved numerically to obtain T_a for different values of the droplet radius a. Then T_a can be inserted into (5.51) and the mass flux $I_T(a)$ can be calculated. This will later be refered to as the *numerical solution*.

The above-described numerical solution requires considerable computational effort. Accordingly, approximate solutions of (5.84) are frequently used in droplet growth theory. Similar to the result of MASON [5.2] the approximate expression

$$I_T(a) \cong \frac{4\pi a(S_\infty - S_a)}{N_M + N_T} \quad , \tag{5.85}$$

where

$$N_M = \frac{1}{\beta_M \cdot F_{M,\infty}} \frac{RT_\infty}{D_\infty M_v p_s(T_\infty)} \quad , \tag{5.86}$$

$$N_T = \frac{1}{\beta_T F_{T,\infty}} \frac{L(T_\infty)}{K_\infty T_\infty} \left[\frac{L(T_\infty)M_v}{RT_\infty} - 1 \right] \quad , \tag{5.87}$$

$$S_a = p_{v,a}/p_s(T_a) \cong \exp\left[\frac{2M_v \sigma_{Lv}(T_\infty)}{R\rho_L(T_\infty)T_\infty} \cdot \frac{1}{a} \right] \left[1 - \frac{3m_N M_v i}{4\pi \rho_L(T_\infty)M_N} \cdot \frac{1}{a^3} \right] \quad , \tag{5.88}$$

$$S_\infty = p_{v,\infty}/p_s(T_\infty) \tag{5.89}$$

can be derived using an approximate expression for the equilibrium vapor pressure $p_s(T)$, as obtained from the Clausius-Clapeyron equation. Using (5.84,70,40), the approximate expression

$$T_a(a) \cong T_\infty + \frac{1}{\beta_T F_{T,\infty}} \cdot \frac{L(T_\infty) \cdot I_T(a)}{4\pi a K_\infty} \tag{5.90}$$

can be derived. Equations (5.85-90) will later be referred to as the *analytical solution*. This solution is restricted to systems in which the temperature is nearly uniform. As mentioned earlier, this condition will be fulfilled approximately for very slow processes and for condensates with small specific latent heat of condensation. Considerable deviations from the numerical solution can be expected at high supersaturations and accordingly high temperature gradients.

5.4.2 Mass and Heat Balance in a Monodispersed Droplet Aerosol

For calculation of the simultaneous growth of randomly located droplets in a supersaturated environment the mutual interaction of the droplets has to be taken into account. A theoretical description of this complicated process can only be performed by the use of simplifying models. FUCHS [5.4] proposed a model that was later called the "cellular model." According to this model, each droplet is surrounded by a sphere with impermeable surface. The volume of this sphere is equal to the system volume divided by the number of droplets. FUCHS pointed out that this model has a nonstationary solution and that a rigorous treatment poses great difficulties. A rough estimate was given. REISS and LAMER [5.57] presented approximate solutions for the case where the production of latent heat is neglected. The solutions are applied to a monodispersed aerosol which is cooled at a constant rate. The non-steady-state calculations of CARSTENS and ZUNG [5.11] dealt with the simultaneous heat and mass transfer in the cellular model thus taking into account the effect of latent heat.

In another model for the calculation of growth processes in droplet populations, it is assumed that steep concentration and temperature gradients only occur in the immediate vicinity of the droplets, and outside of these regions the vapor concentration and temperature profiles between the growing droplets are fairly flat. Accordingly, average bulk parameters of vapor concentration and temperature can be defined. The simultaneous droplet growth can be described by single droplet growth theory with bulk values of supersaturation and temperature that are time dependent due to vapor depletion and production of latent heat. This model of time-dependent bulk parameters was first proposed by REISS [5.58], who calculated the growth process in monodispersed and polydispersed aerosols, neglecting temperature changes due to latent heat production. The results of REISS [5.58] are in agreement with the calculations of REISS and LAMER [5.57], which are based on the cellular model. Based on the model of time-dependent bulk parameters, cloud droplet spectra were calculated by a number of authors, e.g., [5.59-61]. WAGNER and POHL [5.62] used this model to calculate the growth processes at different droplet concentrations in a condensation nucleus counter, taking into account the effect of the production of latent heat.

Both models can only be applied at sufficiently low droplet concentrations, when the mean distance between neighboring droplets is large compared to the droplet radius. If the distance between two droplets is of the order of the droplet radius, an interaction of the density and temperature profiles around the droplets occurs, as described by CARSTENS et al. [5.63]. In this case the droplets do not grow independently and the process cannot be reduced to a single droplet growth. However, for sufficiently low droplet concentrations, an interaction between two droplets is unlikely, and furthermore, the interaction time is short, as pointed out by WILLIAMS and CARSTENS [5.64]. Accordingly, a monodispersed droplet aerosol remains monodispersed during the growth process. REISS [5.58] estimated an upper concentration limit of 10^6 cm^{-3} for droplet radii smaller than 5 μm. However, the assumption of large droplet distances causes no serious limitation of droplet concentration, because at high concentrations, droplet growth terminates at small droplet sizes, so that the ratio of droplet radius and mean droplet distance usually remains small.

The model of time-dependent bulk parameters has the advantage that a random space distribution of the droplets may be considered whereas, strictly speaking, in the cellular model equidistant droplets are supposed. Furthermore, the calculations are much simpler than in the cellular model, because the usual quasi-steady-state equations can be used. However, it is a disadvantage of the model of time-dependent bulk parameters that it may not be applicable at too *small* droplet concentrations. As MASON [5.65] pointed out, transient inhomogeneities in the density and temperature field may occur due to the fast changes of vapor density and temperature at the droplet surfaces. This effect will be more pronounced at large distances between neighboring drops. The actual limits of applicability of the above-described models have not been established. The validity of the approximations must be checked experimentally.

According to the assumed experimental conditions, the droplet aerosol is monodispersed, and size and concentration of the growing droplets are small. Consequently, the mean distance between neighboring droplets is large compared to the droplet radius, and the total volume of the droplets can be neglected compared to the system volume. The vapor concentration and temperature profiles between the growing droplets will be fairly flat with steep concentration and temperature gradients occurring only in the immediate vicinity of the droplets. Accordingly, the model of time-dependent bulk parameters will be applied and the simultaneous droplet growth is described by single droplet growth theory, where time-dependent bulk values of vapor concentration $\rho_{v,\infty}$, vapor pressure $p_{v,\infty}$, temperature T_∞, etc. are used.

The calculation of the above-mentioned bulk parameters is based on the balance of mass and heat in a volume V of the system that is occupied by a particular amount of gas and condensate. μ_g and μ_v are the masses of gas and vapor and N_d is the number of droplets in the volume V. $\rho_{v,\infty} = \mu_v/V$ is the bulk vapor concentration and $C = N_d/V$ is the droplet concentration. μ_g, N_d and the total pressure p_0 are constant during the droplet growth processes. However, μ_v and hence $\rho_{v,\infty}$ will change due to vapor depletion, the bulk temperature T_∞ and hence V and C changing due to the production of latent heat. The ratio T_∞/V is approximately constant, because the changes of μ_v will be small compared to μ_g.

The conservation of the total mass of the condensate in the volume V during the growth process yields the equation

$$(\mu_v)_0 - \mu_v = \frac{4\pi}{3} \rho_L N_d (a^3 - a_0^3) \quad , \tag{5.91}$$

where ρ_L is the density of the liquid and $(\mu_v)_0$, a_0 are the initial values of μ_v, a, respectively. Remembering that T_∞/V is approximately constant, multiplication of (5.91) with RT_∞/VM_v yields the expression

$$p_{v,\infty} = (p_{v,\infty})_0 - \frac{R(T_\infty)_0}{M_v} \rho_L \frac{4\pi}{3} C_0 (a^3 - a_0^3) \tag{5.92}$$

for the bulk vapor pressure, where $(p_{v,\infty})_0$, $(T_\infty)_0$, and C_0 are the initial values of $p_{v,\infty}$, T_∞, and C, respectively.

The conservation of thermal energy in the volume V during the growth process yields the equation

$$[\mu_g c_{p,g} + (\mu_v)_0 c_{p,v}] [T_\infty - (T_\infty)_0] = \frac{4\pi}{3} \rho_L L(T_a) N_d (a^3 - a_0^3) \quad , \tag{5.93}$$

where $c_{p,g}$ and $c_{p,v}$ are the specific heat capacities at constant pressure for gas and vapor, respectively. The term $(\mu_v)_0 c_{p,v}$ in (5.93) is only approximately correct, because μ_v as well as the bulk vapor pressure $p_{v,\infty}$ will change during the growth process. Furthermore, the term $\mu_g c_{p,g}$ in (5.93) is not strictly correct, because the total pressure $p_0 = p_{g,\infty} + p_{v,\infty}$ is constant, whereas the bulk gas pressure $p_{g,\infty}$ will change slightly due to changes of the bulk vapor pressure $p_{v,\infty}$. However, the errors caused by these approximations are negligible. From (5.93) the expression

$$T_\infty = (T_\infty)_0 + \frac{\rho_L L(T_a)}{(\rho_{g,\infty})_0 c_{p,g} + (\rho_{v,\infty})_0 c_{p,v}} \frac{4\pi}{3} C_0 (a^3 - a_0^3) \tag{5.94}$$

can be derived. Equations (5.92,94) describe the dependence of the bulk parameters $p_{v,\infty}$ and T_∞ on the droplet radius a. By inserting these equations into the expressions for the mass and heat flux, the fluxes to a single droplet of the considered droplet aerosol can be obtained, and the mutual interaction of the droplets is approximately taken into account. It should be mentioned that (5.92,94) are somewhat different from the equations (8) and (6) of WAGNER and POHL [5.62]. These differences occur because the equations of [5.62] do not take into account the change of V during the growth process and the heat capacity of the vapor. Numerical calculations have shown, however, that these differences are very small.

5.4.3 Calculation of Droplet Growth and Evaporation

The quasi-steady-state growth rate of the droplets can now be calculated according to

$$\frac{dm_D}{dt} = I_T \quad , \tag{5.95}$$

where $m_D = 4\pi a^3 \rho_L/3$ is the mass of the droplet. Integration of (5.95) yields the growth time t as a function of the droplet radius a:

$$t(a) = \int_{a_0}^{a} \frac{4\pi\rho_L r^2}{I_T(r)}\, dr \quad , \tag{5.96}$$

where a_0 is the initial droplet radius. After inserting the bulk parameters according to (5.92,94) and calculating the droplet temperature T_a and the mass flux I_T by means of the numerical or the analytical solution, the integral in (5.96) can be evaluated numerically and the growth time t(a) can be obtained for various droplet radii a. Numerical inversion of the function t(a) finally yields the droplet radius a(t) as a function of time and the droplet growth calculation is completed. The bulk parameters $p_{V,\infty}$, T_∞ and $S_\infty = p_{V,\infty}/p_S(T_\infty)$ can be obtained as functions of time by inserting the droplet radius a(t) into (5.92,94). Furthermore, the droplet temperature T_a can be calculated as a function of time.

A considerable simplification of the numerical calculations can be achieved by the assumptions that the temperature in the system is nearly uniform, that the droplets are in the continuum regime, and that the droplet concentration is sufficiently small, so that the mutual interactions of the droplets can be neglected. In this case the analytical solution for the mass flux in the continuum regime can be inserted into (5.96) with constant bulk parameters. The integration of (5.96) can be performed in closed analytical form and the well-known quadratic growth law is obtained.

Figures 5.1,2 show curves resulting from numerical calculation for water vapor in air. It can be seen that the droplet temperature T_a remains approximately constant during the growth process. The bulk temperature T_∞ approaches a limiting value as the bulk vapor saturation ratio S_∞ approaches the value 1. During the first stages of droplet growth, a considerable difference between droplet and ambient temperature can be observed. As mentioned earlier, substantial temperature gradients only exist in the vicinity of the droplets. Hence the temperature gradient over

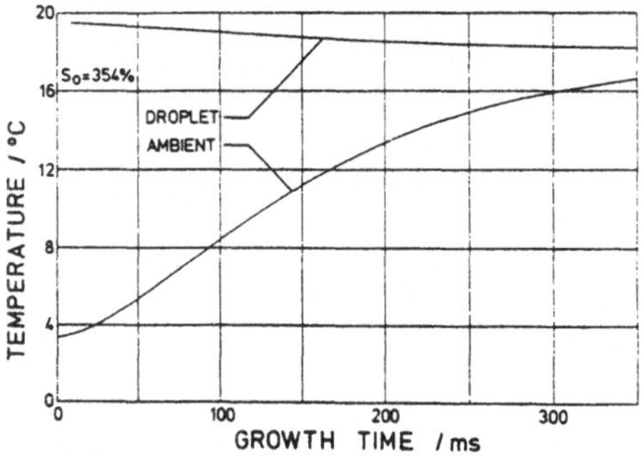

Fig. 5.1. Droplet temperature and ambient gas temperature calculated as functions of time during the growth process

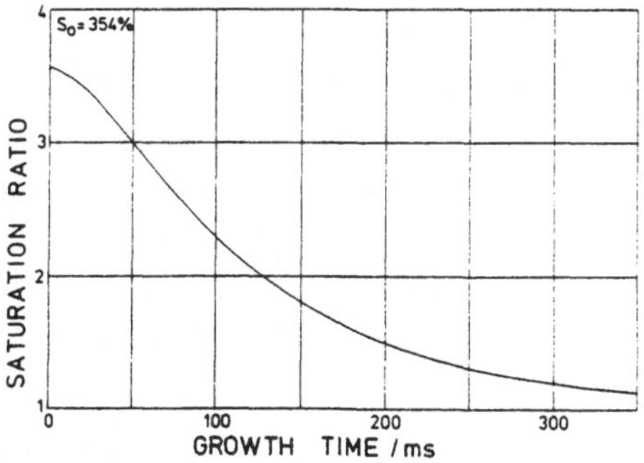

Fig. 5.2. Ambient saturation ratio calculated as a function of time during the growth process

some 10 μm in the neighborhood of the droplets will be extremely high. This explains the importance of the above-mentioned mutual interactions of mass and heat flux.

The theory can be tested by means of comparison between experimental and theoretical droplet growth curves. Theoretical growth curves and a comparison with experimental data will be presented in Sect.5.6. Direct measurements of droplet temperature, ambient vapor pressure and vapor saturation ratio during the growth process are difficult and have not yet been performed.

5.5 Experimental Results

Experimental investigations of droplet growth and evaporation have been performed by many authors. A number of different methods has been applied in order to obtain well-defined vapor pressures and thus saturation ratios in the measuring system. Furthermore, the observation of the droplets was based on various different principles.

5.5.1 Measurements of Particle Evaporation

In connection with measurements of the electron charge, NESTLE [5.66] investigated the evaporation of Hg droplets in a vapor-saturated gas. Charged submicron Hg droplets were injected into a Millikan condenser. Rise and fall times of individual droplets were measured in immediate succession and from these measurements the droplet size was determined. The droplet mass was found to vary approximately linearly with time in agreement with continuum droplet growth theory in connection with Kelvin's formula. Using published Hg vapor pressure data, the diffusivity of Hg vapor in CO_2, N_2, and Ar was estimated.

In a series of papers, BRADLEY and his associates [5.32,67,68] described measurements on the evaporation of drops in a vessel with vapor-absorbing walls. A single droplet with a diameter of the order of 1 mm was suspended on a quartz microbalance and the mass change of the droplet was measured. A number of organic liquids with very low volatility was considered. It was found that the droplet surface varied approximately linearly with time in agreement with continuum droplet growth theory. After the corresponding vapor pressures were measured by means of an independent technique, the vapor diffusivities could be determined from the evaporation rates. Using very low total pressures (down to less than 1 Torr), deviations from continuum droplet growth theory were observed and agreement with FUCHS' [5.4] semiempirical formula was obtained for mass accomodation coefficients close to unity. Similar observations were mady by MONCHICK and REISS [5.69], who considered smaller droplets at somewhat higher pressures in a Millikan condenser with vapor-absorbing walls. LANGSTROTH et al. [5.70] studied the evaporation of single droplets with radii of about 1 mm, suspended in a chamber with vapor-absorbing walls at atmospheric pressure. The droplet radii were measured by means of a microscope. Liquids with somewhat higher volatility, including water, were considered. The experimental evaporation rates were compared with the continuum theory of mass and heat transfer. Using published data for diffusion coefficients and vapor pressures, approximate agreement of theory and experiment was obtained. However, mechanically suspending the liquid droplet might have caused some systematic experimental error, particularly for liquids with higher volatility.

In order to avoid the mechanical suspension of the droplets, a new experimental technique was developed by DAVIS and CHORBAJIAN [5.71], and CHANG and DAVIS [5.72]. A submicron charged droplet with low volatility was suspended in a chamber with zero bulk vapor pressure by means of an inhomogeneous electric field. The droplet size was determined by means of a best-fit procedure between experimental and theoretical light scattering intensities. At atmospheric pressure, a linear time dependence of the droplet surface was observed. Approximate values for vapor pressures and diffusivities were determined from the evaporation data.

Experimental studies of droplet evaporation for low-volatility liquids are hampered by the insufficient knowledge of vapor pressures and diffusivities. Furthermore, significant errors can occur due to small amounts of high-volatility impurities. In order to obtain reliable vapor pressure data and diffusivities for low-volatility liquids, DAVIS and RAY [5.73] developed a method for calculating diffusivity and vapor pressure from continuum droplet evaporation rates, measured in three different carrier gases (He, N_2, and CO_2). Subsequently, this method was improved by RAY et al. [5.74]. Based upon this method, DAVIS and RAY [5.46] performed measurements of droplet evaporation in the transition regime by varying the total gas pressure from 0.08 bar to atmospheric pressure. Good agreement with the theory of LOYALKA [5.42] was obtained.

5.5.2 Measurements of Particle Growth

In the above-mentioned experiments, droplet *evaporation* was investigated. The bulk vapor pressure was either equal to the saturation vapor pressure or equal to zero. For many applications, however, droplet *growth* processes are important. DENNIS [5.75] investigated the growth rates of aqueous solution droplets in an air stream with controlled temperature and relative humidity. Various relative humidities over a range from 73 to 97% were obtained by mixing of a dry and a saturated air stream with well-defined flow rates (divided flow method). A mechanically suspended solution droplet with a diameter of about 1 mm was observed by means of a microscope. The measured growth rates were compared with continuum droplet growth theory taking into account the production of latent heat, the ventilation, and the lowering of the vapor pressure above a solution. Fair agreement was obtained, the experimental growth rate being high by about 30% for some solutions.

For measurements of the growth rate of *pure* liquid droplets, supersaturation must be achieved in the measuring system. Several methods have been used for this purpose. Defined supersaturations can be obtained in static downward diffusion chambers, as described by LANGSDORF [5.76]. Studies of diffusional droplet growth in downward diffusion chambers have been performed by many authors [5.77-80]. As pointed out by VIETTI and FASTOOK [5.81], these studies are complicated by the fact that "the nuclei must be introduced into a saturated environment, where in the time interval allowed for turbulence to subside, the droplets have grown a considerable amount." For measurements of homogeneous nucleation, KATZ and OSTERMIER [5.82] used a static upward diffusion chamber.

COULIER [5.83] used the principle of adiabatic expansion in order to obtain supersaturated vapor. AITKEN [5.84] constructed the first condensation nuclei counter. In the measuring chamber an adiabatic expansion was achieved by moving of a piston (volume-defined expansion chamber). For observing tracks of ionizing particles, C.T.R. WILSON [5.85] designed a volume-defined expansion chamber. Later, C.T.R. WILSON [5.86] constructed a chamber where the expansion was initiated by opening of a valve and thus connecting the chamber to a low-pressure system (pressure-defined expansion chamber). Some important features of expansion chambers were reviewed by J.G. WILSON [5.87]. METTENBURG et al. [5.88] developed an expansion chamber for measurements of condensation nuclei concentrations and droplet growth. This chamber was later improved by KASSNER et al. [5.89].

Expansion chambers have the advantage that supersaturations can be achieved nearly instantaneously, whereas static diffusion chambers work with time-independent supersaturations. Accordingly, droplet growth studies at relatively high supersaturations and growth rates were mostly performed by means of expansion chambers. On the other hand, static diffusion chambers are useful for obtaining low supersaturations, as occuring in the atmosphere. For a correct interpretation of expansion chamber experiments, knowledge of the thermodynamic parameters in the measuring chamber before, during, and after the expansion process is required. In most experiments the considered system is vapor saturated at a known temperature before the expansion occurs. Thereby temperature, vapor pressure and total gas pressure in the expansion chamber are known before the expansion. During and after the expansion, the total gas pressure can be measured. Sufficiently accurate direct measurements of other thermodynamic parameters at the end of the expansion appear to be extremely difficult. These parameters can be calculated from the measured total pressure drop if certain assumptions are fulfilled. Usually it is assumed that the expansion is dry adiabatic, which means that negligible vapor and heat exchange with the chamber walls takes place and vapor depletion and production of latent heat due to droplet growth can be disregarded during the expansion period. The expansion is usually considered as quasistatic, so that the well-known thermodynamic relations for adiabatic processes can be applied. Furthermore, it is assumed that no decomposition of the gas mixture takes place during the expansion period. These assumptions are non trivial, particularly for the case of a pressure-defined expansion chamber.

In order to investigate the thermodynamic processes in expansion chambers, ISRAEL and NIX [5.90,91] measured pressure and temperature in the pressure-defined expansion chamber of a Pollak condensation nuclei counter using fast pressure and temperature transducers. Expansion times of the order of 100 ms were found. Measurements with dry, as well as with moist, air resulted in a temperature decrease much smaller than expected for an adiabatic expansion. According to ISRAEL and NIX, this is caused by an irreversibility of the expansion. The experimental results [5.90,91] were confirmed by SEMONIN and HAYES [5.92]. By means of light extinction measurements ISRAEL and NIX observed the onset of water droplet growth 7 ms after the start of the expansion. The actual saturation ratio in the expansion chamber of the POLLAK counter was found by ISRAEL and NIX to be smaller than 102% during the whole expansion process, whereas a saturation ratio of 270% would be expected according to dry-adiabatic calculations [5.93]. NIX [5.93,94] concluded from the above results that the experimental droplet growth rate exceeds the theoretical prediction by a factor of about 100. KASSNER et al. [5.95,96] discussed the difficulties in connection with temperature measurement using fine-wire thermometers in gases. The finite heat capacity of the wire and the steady heat production can cause erroneous temperature measurements, particularly during fast temperature changes. Finally, it is concluded [5.96] that "in all probability it (the POLLAK counter) is sufficiently adiabatic."

In order to measure droplet growth rates under *reversible* thermodynamic conditions, NIX [5.93,94,97] developed a new experimental technique. Instead of obtaining supersaturation by means of one single expansion, NIX achieved periodic changes of the saturation ratio by applying periodic pressure oscillations in a chamber with vapor-saturated air. Sinusoidal pressure oscillations with frequencies between 1 and 60 Hz and pressure amplitudes between 9 and 13 Torr were obtained by means of an electronic wave generator in connection with an electrodynamic pressure transducer. Corresponding peak saturation ratios between 104 and 108% were calculated. The periodic growth and evaporation of single water droplets was optically observed. In order to achieve a sufficient time resolution, the considered droplets were kept on stationary elliptical orbits by introducing a slight asymmetry of the flow conditions in the measuring chamber. By evaluating photographs of the droplets on their elliptical orbits, NIX [5.94] found that droplet growth only occurs during the periods of *increasing* supersaturation. Based on this observation NIX [5.94] concluded that droplet growth is determined by a "slow" diffusion process, which depends on S and a "very fast" gas kinetic process depending on dS/dt. Later, NIX [5.93] applied a light scattering technique for observation of the droplets and found a sinusoidal droplet size change, the growth process occuring during nearly the *whole* period of supersaturation. However, no quantitative comparison with droplet growth theory was performed.

OWE BERG and GEORGE [5.98] investigated water droplet growth in a volume-defined expansion chamber with an expansion time of about 200 ms. The droplets were observed during the expansion by means of a high-speed film camera. However, no droplet growth could be detected. The droplet concentration was found to increase during the observation time, the droplet aerosol being monodispersed at all times. OWE BERG and GEORGE concluded that "the growth of new droplets was too rapid to be resolved by the camera" and thus "50 μsec is a generous upper limit for the time of formation of condensate droplets 4 or 5 μ in diameter." As the known theories cannot explain these high growth rates, OWE BERG and GEORGE presume that the droplets "are formed by a chain-reaction type mechanism," the condensation nuclei serving as "chain initiators." However, KASSNER et al. [5.95] point out that "it still requires the order of one millisec to increase the radius of a drop by one micron under the fastest conceivable physical mechanism." SCHUSTER et al. [5.99] show that, due to the small sampling volume in the chamber used by OWE BERG and GEORGE, strong fluctuations of the observed droplet number will occur causing large statistical errors. Furthermore, NIX [5.94] suspects that because of an insufficient resolution of the optical system used, the droplets will have caused a diffraction disc with a constant diameter between 3 and 4 μm. This would explain why actual droplet growth was not observed by OWE BERG and GEORGE.

In order to test the droplet growth theory, quantitative measurements of the droplet size must be performed. COHEN [5.100] first suggested a measuring technique based on illuminating the growing droplets and monitoring the intensity of the light scattered by the droplets under a particular fixed scattering angle during the growth process. COHEN considered water droplets in a thermal diffusion cloud nuclei chamber at saturation ratios between 100.4 and 101% and measured the scattered light intensity under a scattering angle of 90^0. Because of the complexity of the obtained light scattering curves and the relatively low scattering intensities, a unique correspondence to the theoretical curves was not established. Thus the absolute size of the growing droplets was not determined and no droplet growth curves were obtained [5.101].

A similar experimental technique was used by VIETTI and SCHUSTER [5.102,103] for measurements of water droplet growth in an expansion chamber at saturation ratios ranging from about 129 to 345%. The saturation ratios were calculated assuming dry-adiabatic conditions. The chosen expansion chamber was a duplicate of that used by ALLARD and KASSNER [5.104] with a chamber volume of about 10 l and expansion times of the order of 100 ms. Reevaporation nuclei from a previous homogeneous nucleation were used as condensation nuclei. The nuclei concentrations were of the order of 10 cm^{-3}. At these very low concentrations the dry-adiabatic assumption appears to be justified. VIETTI and SCHUSTER chose a forward scattering angle of 30^{0} and accordingly obtained less complicated experimental light scattering curves and higher scattering intensities than COHEN [5.100]. The correspondence between experimental and theoretical light scattering extrema was obtained by identification of the first Mie peak. As pointed out by CARSTENS and CARTER [5.105], "the problem of initial conditions is important in using this technique. There is no guarantee that the first *observed* peak is actually the first Mie peak. This may be due to the possibility that the first peak, which is weak anyway, is not resolved. It may also be that the size distribution is just broad enough at the first peak, that it is washed out." In the experiments of VIETTI and SCHUSTER [5.102,103] a "very monodisperse population of droplets" was observed. After the one-to-one correspondence between experimental and theoretical light scattering extrema was established, the droplet size at particular times during the growth process was determined from the position of the experimental light scattering extrema. The droplet growth was measured over a range of radii between 0.45 and 10 μm. Fair agreement with several droplet growth theories was found, but mass accomodation coefficients as small as 0.004 and simultaneously thermal accomodation coefficients of 0.1 and less had to be assumed.

Using the same experimental technique, CARSTENS et al. [5.106] measured water droplet growth in argon and found that a good theoretical fit to the experimental data can be obtained by proper adjustment of the accomodation coefficients. It is emphasized that "a key step in the interpretation of the data lies in the identification of a specific Mie scattering peak." Subsequently, CARTER and CARSTENS [5.107] obtained "much better agreement" with the "conventional meteorological theory of drop growth" over a range of saturation ratios from 130 to 360% assuming a thermal accomodation coefficient of unity and a mass accomodation coefficient of 0.023.

WAGNER [5.108] applied a similar experimental technique for quantitative measurement of water droplet growth. However, a smaller chamber volume of about 70 cm^3 was chosen and expansion times as short as 5 - 7 ms were thus achieved. Using different scattering angles in the range from 15^{0} to 165^{0}, a unique correspondence between experimental and theoretical light scattering peaks was established. Experimental droplet growth curves were obtained but no comparison with droplet growth theory was performed.

In order to extend the range of supersaturations covered by the above-mentioned expansion chamber experiments, GOLLUB et al. [5.78,79] studied water droplet growth in a static downward diffusion chamber at saturation ratios not higher than 105%. Condensation nuclei were introduced into the chamber by means of a slow vertical downward flow of air whose influence on the static supersaturation and temperature profiles in the chamber was negligible. The droplets, condensing on the condensation nuclei, fall through a sequence of known environments at a speed given by the sum of the mean flow speed and the terminal settling velocity of the droplets. For steady-state conditions, the droplet radius and the corresponding fall velocity will only be functions of the height in the chamber. GOLLUB et al. [5.78] described a technique for measuring the fall velocity of the droplets using optical heterodyne spectroscopy. Their method is applicable for droplets of radius larger than about 2 μm, because in this size range the influence of Brownian motion on the heterodyne spectrum is negligible. From measurements of the fall velocity at various heights, the droplet radius can be calculated as a function of height. Qualitative agreement with the theory of FUKUTA and WALTER [5.35] was obtained for a mass accomodation coefficient of 0.02. After improving the experimental design, GOLLUB et al. [5.79] obtained agreement with the theory of FUKUTA and WALTER for saturation ratios between 102 and 105%, if mass and thermal accomodation coefficients are related by a particular equation, but $\alpha = \beta = 1.0$ is not definitely excluded. There appears to be a definite contrast to the results of VIETTI and SCHUSTER [5.102,103]. However, strong deviations from theory were observed for saturation ratios smaller than 101.5%. It should be mentioned that some uncertainty of the actual supersaturations is introduced by the presence of the cylindrical walls of the chamber.

Up to this point, most experiments, e.g., VIETTI and SCHUSTER [5.102,103], were performed with extremely low droplet concentrations and thus the mutual influence of the growing droplets was assumed to be negligible. In order to study the influence of droplet concentration on the droplet growth process, WAGNER [5.109] measured water droplet growth in an expansion chamber at two different saturation ratios, varying the droplet concentration over a range from 5×10^3 to 500×10^3 cm^{-3}. Using the same technique as described above, the droplet size at particular times during the growth process was determined from the *position* of the experimental light scattering extrema. But furthermore, the droplet concentration was independently obtained from the *height* of the experimental light scattering maxima. It was found that the droplet growth is nearly unaffected by droplet concentration during the initial growth stages. However, later in the growth process, a stoppage of growth was observed depending on the actual droplet concentration. This behavior was qualitatively explained by the effects of vapor depletion and production of latent heat during droplet growth, but no quantitative comparison with corresponding droplet growth calculations was performed.

For calculation of droplet growth in droplet aerosols, WAGNER and POHL [5.62] as a first approximation used an analytical continuum droplet growth expression [5.2] without transitional corrections. The combined effects of vapor depletion and production of latent heat in the droplet aerosol were taken into account using the model of time-dependent bulk parameters (Sect.5.4.2). Droplet radius, bulk temperature, and bulk saturation ratio were calculated as functions of time for different initial saturation ratios and droplet concentrations. In qualitative agreement with the experimental results of WAGNER [5.109], the calculated droplet growth curves turned out to be insensitive with respect to the droplet concentration during the initial growth stages. Furthermore, a stoppage of growth is predicted depending on the actual droplet concentration. When the droplet radius approaches its final value, the calculated bulk saturation ratio approaches 100%. A quantitative comparison of experimental and theoretical growth curves was performed by WAGNER [5.110]. The experimental growth rate was found to be significantly smaller than theoretically predicted. As a possible explanation WAGNER suggested insufficient humidification of the condensation nuclei aerosol. After assuming a certain value less than 100% for the relative humidity of the nuclei aerosol, good agreement with the experimental growth curves was obtained. The dependence of the droplet growth rate on the droplet concentration was correctly described by the theory used.

VIETTI and FASTOOK [5.111] reported measurements performed in an expansion chamber modified to produce small supersaturations. Water droplet growth was observed at saturation ratios below 110% and droplet concentrations of no more than 10^3 cm^{-3}. In agreement with WAGNER [5.109] it was found that the growth process stopped at a certain terminal size. From comparison with theory, VIETTI and FASTOOK [5.111] concluded that "theory agrees reasonably well with experiment during growth of droplets, but does not predict the stoppage of growth."

Using a volume-defined expansion chamber with a volume of 1 m^3, VIETTI and FASTOOK [5.81] measured water droplet growth in air for peak saturation ratios ranging from 102 to 105%. Because of the large size of the chamber, the central portion of the chamber volume remains adiabatic for as long as 30 s. The expansion times were of the order of 300 ms. Droplet concentrations of approximately 500 cm^{-3} were chosen for all experiments. Again, stoppage of growth was observed at particular terminal droplet sizes. Based on the theory of FUKUTA and WALTER [5.35] and taking into account vapor depletion and production of latent heat, droplet growth curves were calculated. The best overall agreement with the experimental data was obtained if both accomodation coefficients were chosen to be unity. This is in contrast to the results obtained by VIETTI and SCHUSTER [5.102,103] for higher supersaturations. VIETTI and FASTOOK [5.81] conclude that the theory of FUKUTA and WALTER is valid for atmospheric conditions. At larger supersaturations, however, deviations occur and the experiments at larger supersaturations are regarded as "not similar." According to VIETTI and FASTOOK, this "places the larger supersaturation cloud chamber data as not applicable" with respect to atmospheric conditions.

The dependence of the droplet growth process on the carrier gas was studied by VIETTI and FASTOOK [5.112]. Water droplet growth measurements in carbon dioxide were performed for saturation ratios between 101.7 and 103.5%. It turned out that there is "no specific value for the sticking coefficient which provides overall agreement during the later stages of the observed growing period." For droplets smaller than 2 μm radius, the mass accomodation coefficient α_M assumes values between 1 and 0.1. However, above 2 μm radius, α_M must be continually diminished to a value around 0.04. VIETTI and FASTOOK suggest that the dehydration of bicarbonate is the principal mechanism for the observed reduction of droplet growth rate. In order to investigate the influence of the condensation nuclei on the growth process, VIETTI [5.113] produced different NaCl particles by spraying aqueous salt solutions of different concentrations. It was found that $\alpha_M = 0.1$ for pure water and $\alpha_M = 0.02$ for the case when 10^{-17} mol salt is present in each droplet.

Using a process-controlled expansion chamber, WAGNER and POHL [5.114,115] measured water droplet growth rates for different initial saturation ratios. Systematic deviations from the droplet growth theory used were observed, stronger deviations occuring at higher saturation ratios. A fit of the theoretical curves with the experimental data can be achieved by proper choice of the assumed humidification of the nuclei aerosol. However, smaller values of the humidification had to be assumed for higher saturation ratios. On the other hand, the measuring system was designed so that the humidification of the nuclei aerosol is definitely independent of the chosen expansion ratio. This indicates that the dependence of the droplet growth rate on the supersaturation is not properly described by the applied theoretical model.

It appears from the above-mentioned experiments that the deviations between theory and experiment are increasingly important for higher saturation ratios and growth rates. Furthermore, there is still much uncertainty about the actual values of the accomodation coefficients and their possible dependence on other physical parameters.

5.6 Comparison of Numerical Growth Calculations with Recent Expansion Chamber Experiments

The experimental results mentioned in Sect.5.5 were generally compared with droplet growth calculations based on the phenomenological equations (5.17,22) (Fick's and Fourier's law), and a linearized expression for the vapor pressure was used. The observed deviations indicate that the zeroth-order theoretical model does not provide a satisfactory description of droplet growth. In the present section the first-order theoretical model, described in Sects.5.2-4 is applied and the results of numerical first-order calculations for the system water vapor in air are described. Droplet growth curves are presented for three different initial saturation ratios and the influence of various physical parameters on the growth process is demonstrated. Thereby some additional information about the physical processes occuring during droplet growth is obtained.

The calculated growth curves are compared with experimental data obtained by means of a process-controlled expansion chamber recently developed at the University of Vienna [5.116,117]. The pressure-defined thermostated expansion chamber EXP (Fig.5.3) has a volume of approximately 130 cm^3. For observation, the droplets, growing in EXP, are illuminated by a continuous wave He-Ne laser. The light flux

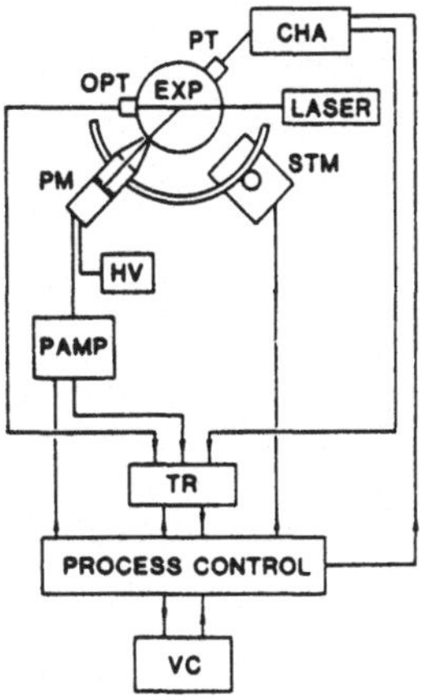

Fig. 5.3. Data acquisition system of the process-controlled expansion chamber

transmitted through the expansion chamber is measured by an optoelement OPT. The light flux scattered by the growing droplets is monitored under selectable scattering angles by a photomultiplier PM in connection with a high-voltage power supply HV and a process-controlled amplifier PAMP. The scattering angle can be varied over a range from 15^{0} to 165^{0} by means of a process-controlled step motor STM, the PAMP being automatically adjusted. The total gas pressure in EXP is measured by a fast, piezoelectric pressure transducer PT in connection with a charge amplifier CHA. Transmitted and scattered light flux and total pressure are recorded simultaneously by a process-controlled digital transient recorder TR and subsequently transferred into the computer and stored. In order to obtain good reproducibility, the measuring process is controlled by an electronic valve control unit VC in connection with five solenoid valves. VC determines the timing of the experiment and interacts with the process-control system.

A typical set of experimental data is presented in Fig.5.4. The upper curve shows the total gas pressure in EXP, the middle curve the transmitted light flux (inverted) and the lower curve the scattered light flux (forward scattering angle 15^{0}) as a function of time. It can be seen that the expansion process has a duration of 5 - 7 ms and is completed before the droplets grow to detectable sizes. Accordingly, a dry-adiabatic expansion is assumed and initial temperature and saturation ratio at the end of the expansion are calculated from the measured total pressure drop. The experimental scattered light flux versus time curve (Fig.5.4, lower curve) shows series of maxima and minima in good agreement with the corresponding theoretical scattered light flux versus size curve (Fig.5.5) calculated by means of Mie theory [5.118]. This indicates that the droplets are quite monodispersed during the growth process. After measuring the scattered light flux for a series of different scattering angles [5.119,120], a one-to-one correspondence between experimental and theoretical light scattering extrema can be established uniquely. This allows a simultaneous and independent quantitative determination of droplet radius and droplet concentration at specific times during the growth process.

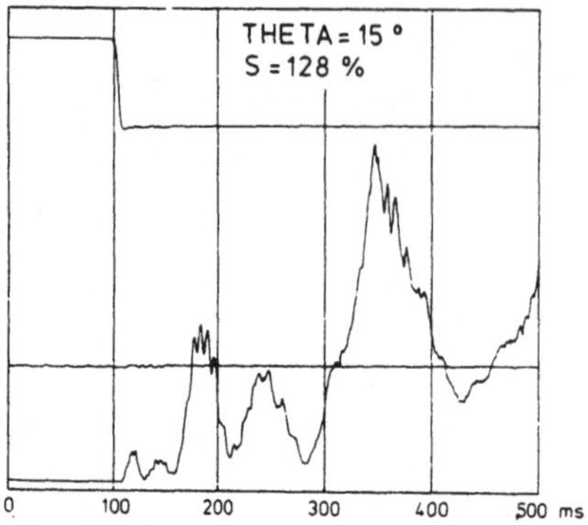

Fig.5.4

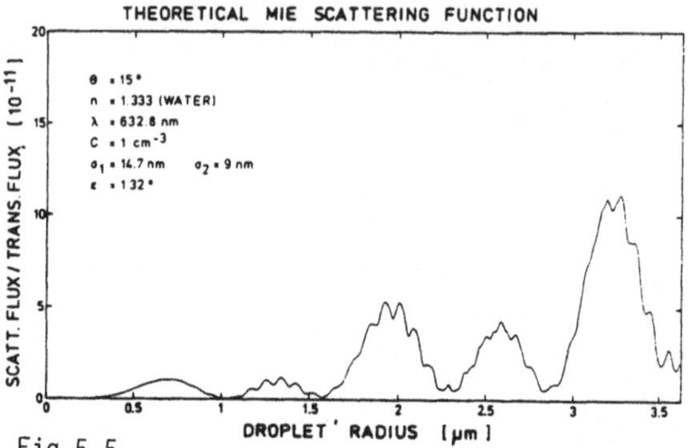

Fig.5.5

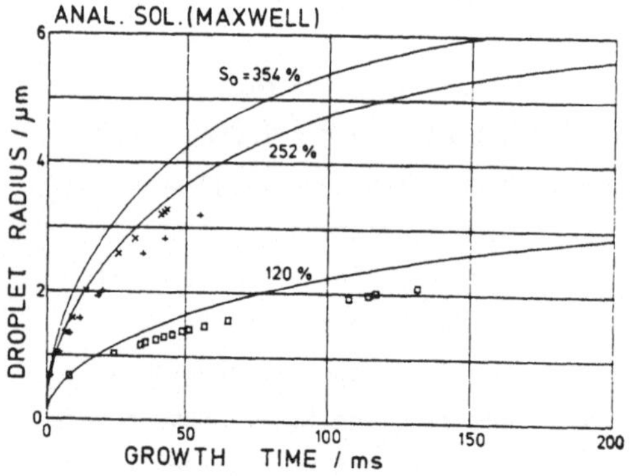

Fig. 5.4. Typical set of experimental data as functions of time. Upper curve: total pressure. Middle curve: transmitted light flux (inverted). Lower curve: scattered light flux

Fig. 5.5. Corresponding theoretical scattered light flux vs. size curve calculated according to Mie theory

Fig. 5.6. Theoretical growth curves for S_0 = 120, 252, 354%, analytical solution, no corrections. The symbols (□,+,×) are experimental data [5.121,122] for S_0 = 120, 252, and 354%, respectively

From the position of the experimental light scattering extrema, the droplet size at specific times can be obtained. On the other hand, from the height of the experimental maxima the droplet concentration can be evaluated. The light extinction occuring at high droplet concentrations in the expansion chamber is taken into account by normalizing the experimental scattered light flux relative to the laser light flux transmitted through the expansion chamber.

Droplet growth measurements [5.117,121,122] were performed for the system water vapor in air at initial saturation ratios 120, 252 and 354% and at a droplet concentration of 4.6×10^3 cm^{-3}. The experimentally obtained droplet growth data were compared with various theoretical predictions. Figure 5.6 shows a comparison with droplet growth curves calculated by means of (5.85-90) (analytical solution). The correction factors β_M, β_T, F_M, and F_T were not applied and thereby similar equations, as originally derived by MAXWELL [5.16], were obtained. As can be seen from Fig.5.6, the experimental droplet growth rates are significantly smaller than theoretically predicted by the uncorrected analytical solution over the considered range of saturation ratios. Some improvement of the theoretical fit can be achieved by introducing the transitional correction factors β_M and β_T in (5.85-90). Figure 5.7 shows the numerical results together with the corresponding experimental data for the initial saturation ratio of 354%. The upper curve is based on the analytical solution (5.85-90) with no transitional correction included. The middle and lower curve corresponds to the semiempirical expression of SMIRNOV [5.41], (5.73,74), and FUKUTA and WALTER [5.35], (5.71,72), respectively. The calculations according to DAHNEKE [5.37] and SMIRNOV yielded very similar results, which could not therefore be indicated separately. As can be seen from Fig.5.7, the experimental growth rate is still significantly smaller than theoretically predicted. The differences between the various transitional corrections are rather small in the range of Knudsen numbers considered.

For the above calculations the mass and thermal accomodation coefficients α_M, α_T were both chosen to be unity. As suggested by many authors (Sects.5.3.2 and 5.5), a better agreement could be obtained by assuming lower values for the accomodation coefficients. It turns out that by proper choice of α_M the theoretical curves can be forced to agree with the experimental data for small droplet radii. However, significant deviations still occur at larger droplet sizes. In this connection it should be mentioned that the transitional correction is insignificant for larger droplet sizes and accordingly small Knudsen numbers. As mentioned in Sect.5.3.2, the transitional correction becomes less important with increasing droplet radius, whereas the diffusional and thermal corrections are size independent. The relative influence of these corrections is illustrated in Fig.5.8 for an initial saturation ratio of 354%. The upper curve is based on the analytical solution (5.85-90) with no corrections included. The second curve from above corresponds to the transitional correction of SMIRNOV [5.41]. The third curve from above was calculated taking into account the diffusional and thermal correction factors F_M, F_T according to (5.46,39) in addition to the transitional correction factors β_M, β_T. Mass and thermal accomodation coefficients α_M, α_T were both chosen to be unity and the thermal diffusion factor α was chosen to be zero. It can be seen that the diffusional and thermal correction provides a significant improvement of the theoretical fit. This improvement was achieved without adjusting undetermined parameters.

While a satisfactory theoretical fit has been obtained for small droplet radii, significant deviations still occur at larger sizes. These deviations cannot be eliminated by chosing accomodation coefficients less than unity, because as mentioned above, the transitional correction is insignificant for larger droplet sizes. A possible explanation for the remaining deviations could be the fact that the expression (5.85) for the mass flux was derived using an approximate expression for the equilibrium vapor pressure, as obtained from the CLAUSIUS-CLAPEYRON equation. The correct temperature dependence of the equilibrium vapor pressure can be taken into account by numerically solving (5.84) to obtain T_a for different values of

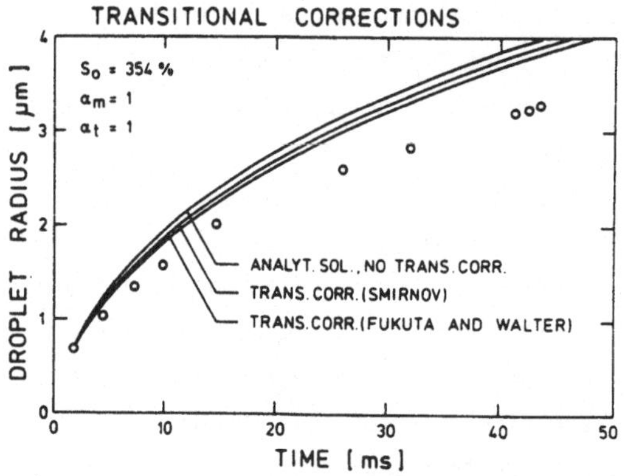

Fig. 5.7. Theoretical growth curves for S_0 = 354%, analytical solution, no corrections (upper curve), transitional corrections according to SMIRNOV [5.41] (middle curve), FUKUTA and WALTER [5.35] (lower curve). Experimental data are indicated

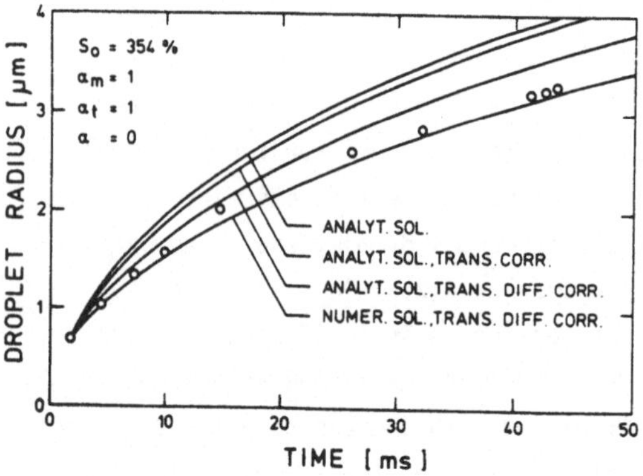

Fig. 5.8. Theoretical growth curves for S_0 = 354%: analytical solution, no corrections (upper curve), transitional correction (second curve from above); transitional diffusional and thermal corrections (third curve from above); numerical solution, all corrections (lower curve). Experimental data are indicated

the droplet radius a (numerical solution). The lower curve in Fig.5.8 was calculated for the initial saturation ratio 354% by means of the numerical solution. The transitional correction according to SMIRNOV [5.41] with $\alpha_M = \alpha_T = 1$ as well as diffusional and thermal corrections with $\alpha = 0$ were applied. Good agreement with the experimental data can be observed. It should be emphasized that the theoretical curves of Fig.5.8 were obtained without adjusting undetermined parameters.

The above-described numerical results were obtained taking into account the effects of vapor depletion and production of latent heat during the growth process (Sect.5.4.2). The influence of these effects on the growth curves is illustrated by means of Fig.5.9. The lower curve in Fig.5.9 was calculated taking into account vapor depletion and latent heat production for the actual droplet concentration of 4.6×10^3 cm^{-3} and using the numerical solution with all corrections applied. This curve is identical to the lower curve in Fig.5.8. The upper curve in Fig.5.9 was calculated similar to the lower curve, but assuming the droplet concentration to be zero and thereby neglecting the mutual influence of the growing droplets. The small differences between the two curves in Fig.5.9 indicate that the droplet concentration was chosen sufficiently small so that the mutual influence of the growing droplets does not cause significant changes of the growth curves for the considered measuring times.

There is some uncertainty about the actual value of the thermal diffusion factor α. As mentioned in Sect.5.2.3, α is probably equal to 0.01 for the mixture of water vapor and air [5.26,28]. Generally, the absolute value of α is assumed to be less than 0.6 (Sect.5.2.2). Figure 5.10 shows the influence of the thermal diffusion factor α on the growth curves. It can be seen that even thermal diffusion factors with unrealistically high absolute values cause only small changes of the growth curves.

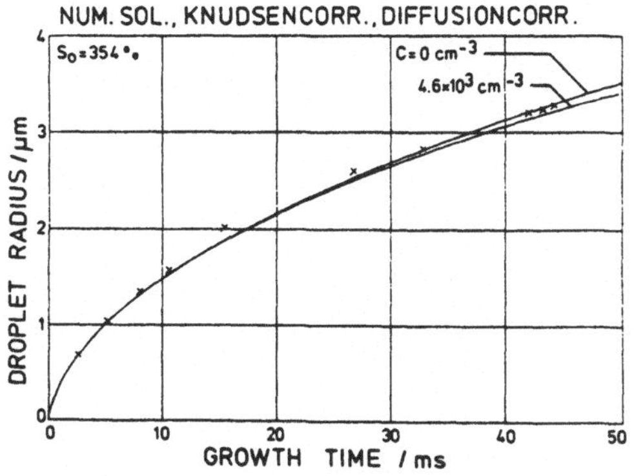

Fig. 5.9. Influence of the number concentration C. Experimental data are indicated

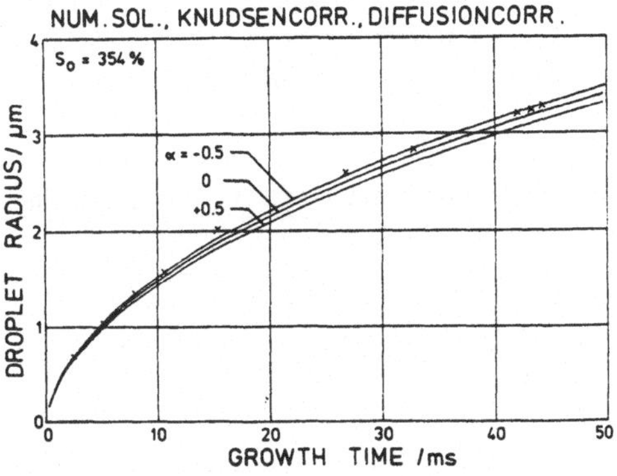

Fig. 5.10. Influence of the thermal diffusion factor α. Experimental data are indicated

The dependence of the theoretical growth curves on the choice of the mass accomodation coefficient α_M for the initial saturation ratio 354% and for a thermal accomodation coefficient $\alpha_T = 1$ is demonstrated in Fig.5.11. The calculations were based on the numerical solution with all corrections included. The best theoretical fit can be observed for $\alpha_M = 1$. For values of the mass accomodation coefficient $\alpha_M < 0.5$, significant deviations between theoretical and experimental growth rates can be observed. Chosing $\alpha_M = 0.03$ causes strong deviations from the experimental data. Accordingly, for the present experimental conditions, the mass accomodation coefficient α_M for water was found to be larger than 0.5. As mentioned earlier, transitional effects and thus accomodation coefficients have a significant influence on the droplet growth process only at sufficiently large Knudsen numbers. Therefore a more precise determination of α_M can only be performed if smaller droplets or lower gas pressures are considered.

In order to test the first-order theory of droplet growth for different initial supersaturations, the calculations were performed for the initial saturation ratios 120, 252, and 354%. Figure 5.12 shows theoretical growth curves calculated by means of the numerical solution with the transitional correction according to SMIRNOV as well as diffusional and thermal corrections included. Both accomodation

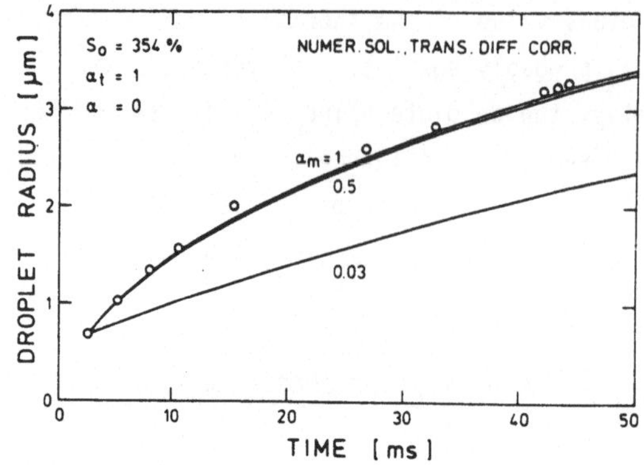

Fig. 5.11. Influence of the mass accomodation coefficient α_M. Experimental data are indicated

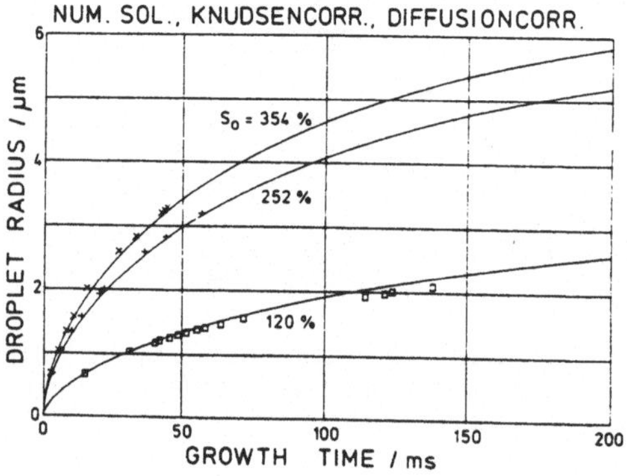

Fig. 5.12. Theoretical growth curves for S_0 = 120, 252, 354%, numerical solution, all corrections applied. The corresponding experimental data are indicated

coefficients were chosen to be unity and the thermal diffusion factor was set equal to zero. Satisfactory agreement with the experimental data can be observed for the range of supersaturations considered without making arbitrary assumptions. This indicates the consistency of the first-order theoretical model.

For the above calculations the initial saturation ratios in the expansion chamber were determined assuming a reversible, quasistatic and dry-adiabatic expansion. The above-mentioned consistency of the theoretical results at different initial supersaturations and thus different expansion ratios suggests that these assumptions are justified for the present experimental conditions. The expansion can only be considered as dry adiabatic if changes of the saturation ratio during the expansion process due to vapor depletion and production of latent heat are negligible. Direct measurements of the saturation ratio during the growth process are difficult, as mentioned in Sect.5.4.3. However, based on the agreement of theoretical and experimental droplet growth rates it can be assumed that the calculated time dependence of the bulk saturation ratio agrees with the actual experimental conditions. Figure 5.2 in Sect.5.4.3 shows the bulk saturation ratio calculated for the present experimental conditions and for the initial saturation ratio 354%

as a function of time during the growth process. The calculations were performed by means of the numerical solution with the Knudsen correction according to SMIRNOV and diffusional and thermal corrections included. It can be seen from Fig.5.2 that the changes of the bulk saturation ratio during the first 10 ms of the growth process are small. As shown in Fig.5.4, the expansion process has a duration of 5 - 7 ms. This further supports the assumption of dry-adiabatic expansion. Accordingly, it can be assumed as a good approximation that for the present experimental conditions the calculated initial saturation ratios actually occur in the expansion chamber at the end of the expansion.

5.7 Conclusions and Outlook

The results described in Sects.5.5,6 allow a number of conclusions to be drawn.

1) The assumption of quasi-steady state provides a good theoretical description of droplet growth processes for most practically important conditions. Accordingly, the duration of transient phenomena seems to be small compared to the typical growth times. This experimental result is in general agreement with most theoretical investigations of nonstationary droplet growth and evaporation (Sect.5.1).

2) Possible effects of incomplete mass or thermal accomodation at the droplet surface are not important for the description of the growth of pure water droplets in air over a range of saturation ratios from 120 to 354% and for Knudsen numbers below about 0.3. The best theoretical fit with the experimental growth rates was obtained by setting the mass as well as the thermal accomodation coefficient equal to unity. As mentioned in Sects.5.3.2. and 5.5, many authors assume mass accomodation coefficients α_M for water of the order of 0.03 and smaller; the value $\alpha_M = 0.036$, as obtained by ALTY and MACKAY [5.50], was used by a number of authors. However, based on the results presented in Fig.5.11, values for the mass accomodation coefficient α_M of the order of 0.03 are definitely excluded for the experimental conditions considered.

3) The first-order mutual interactions of mass and heat flux in the vicinity of the droplets have a significant influence on the droplet growth process. As opposed to the transitional effects, the first-order effects are not size dependent and thus are important even at larger droplet sizes where the transitional effects are already negligible. In previous investigations, the first-order effects have not been taken into account. This is consistent with the fact that most authors report experimental droplet growth rates smaller than theoretically predicted. In many cases the agreement was then achieved by chosing accomodation coefficients less than unity. This might be an explanation why low and very scattered values of the

accomodation coefficients for water can be found in the literature. From the above-described calculations it can be concluded that thermal diffusion and the inverse process, the diffusion-thermo effect, are of minor importance for the droplet growth, at least for the system water vapor in air. This results agrees with an estimate by VIETTI and FASTOOK [5.81]. Furthermore, the calculations showed that the Stefan flow is negligible. However, there are two first-order effects which are important for droplet growth processes. The mass flux is smaller as compared to the usually applied zeroth-order theory because the driving force for the mass flux is grad X_v and not grad ρ_v. These gradients are proportional only for the case of uniform temperature, but high temperature gradients (of the order of 15×10^3 K cm^{-1}) occur in the vicinity of the droplets. Furthermore, the heat flux is retarded because of the kinetic energy, which is carried with the diffusing vapor molecules. Both effects together cause a significant reduction of the theoretical droplet growth rate. For high supersaturations it was found that the analytical solution (5.85-90), which is based on an approximate expression for the saturation vapor pressure, significantly deviates from the experimental data. The numerical solution of (5.84), however, yields a good theoretical fit over the whole range of supersaturations considered. This indicates that using a simplified approximate expression for the equilibrium vapor pressure causes errors at high supersaturations. In most previous investigations, droplet growth was calculated by means of the analytical solution. This might be an explanation for the fact that VIETTI and SCHUSTER [5.102,103] found deviations between theory and experiment for high supersaturations, whereas GOLLUB et al. [5.79], and VIETTI and FASTOOK [5.81] observed good agreement for low supersaturations. It can be concluded that droplet growth for atmospheric conditions at low supersaturations can be described sufficiently accurate by means of the analytical solution (5.85-90). However, it is important to include the diffusional and thermal correction factors F_M and F_T according to (5.46,39) in addition to the transitional correction factors β_M and β_T. After inserting F_M and F_T into the analytical growth law, good agreement with experimental growth rates at low supersaturations is obtained without arbitrary adjustment of undetermined parameters.

In order to obtain more detailed information about growth processes in the transition regime, experiments must be performed over an extended range of Knudsen numbers. This would allow more accurate determinations of the thermal and mass accomodation coefficients. Many authors have measured the growth rate of water droplets in air. It would be particularly important to study other vapor-gas combinations and to investigate the possible influence of the carrier gas on the accomodation coefficients [5.112] and on droplet growth. Furthermore, investigations of the growth of mixed droplets from a mixture of vapors could provide useful information concerning processes occuring in the atmosphere and technological applications. In most experiments the growth processes are considered at sufficiently low particle number concentrations that single droplet growth is observed. Additional information about growth processes in aerosols and the mutual interaction of growing particles can be obtained by measuring growth rates over an extended range of particle concentrations [5.109-111]. Thereby the dependence of the growth rate on the particle number concentration and the actual limits of applicability for the model of time-dependent bulk parameters could be investigated experimentally.

Acknowledgement. The author wishes to thank Professor Dr. O. Preining for many helpful discussions and for the interest he has shown in this work. In addition many useful suggestions by Dr. F.G. Pohl and his help with the numerical calculations are acknowledged.

Thanks are also due to Mrs. H. Kranner for typing and proofing the manuscript. The author gratefully acknowledges the support by the Österreichisches Bundesministerium für Wissenschaft und Forschung, the University of Vienna, The EDV-Zentrum der Universität Wien, Abteilung Prozeßrechenanlage Physik, the Fonds zur Förderung der wissenschaftlichen Forschung in Österreich, Project No. 2429,3481, and the Hochschuljubiläumsstiftung der Stadt Wien.

References

5.1 N.A. Fuchs: *Evaporation and Droplet Growth in Gaseous Media* (Pergamon, London 1959)
5.2 B.J. Mason: *The Physics of Clouds*, 2nd ed. (Clarendon, 1971)
5.3 Y.S. Sedunov: *Physics of Drop Formation in the Atmosphere* (Wiley, New York 1974)
5.4 N. Fuchs: Phys. Z. Sowjet. *6*, 224 (1934)
5.5 N. Frössling: Gerlands Beitr. Geophys. *52*, 170 (1938)
5.6 P. Squires: Aust. J. Sci. Res. *5*, 59 (1952)
5.7 J.R. Brock: J. Colloid Interface Sci. *24*, 344 (1967)
5.8 G. Luchak, G.O. Langstroth: Can. J. Res. *A 28*, 574 (1950)
5.9 H.L. Frisch, F.C. Collins: J. Chem. Phys. *21*, 2158 (1953)
5.10 J.S. Kirkaldy: Can. J. Phys. *36*, 446 (1958)
5.11 J.C. Carstens, J.T. Zung: J. Colloid Interface Sci. *33*, 299 (1970)
5.12 R. Chang, E.J. Davis: J. Colloid Interface Sci. *47*, 65 (1974)
5.13 N. Nix, N. Fukuta: J. Chem. Phys. *58*, 1735 (1973)
5.14 S. Twomey: *Atmospheric Aerosols*, Developments in Atmospheric Science, Vol. 7 (Elsevier, Amsterdam 1977)
5.15 J.R. Brock: J. Colloid Interface Sci. *23*, 286 (1967)
5.16 J.C. Maxwell: "Diffusion", Encyclopedia Britannica *2*, 82 (1877); reprinted in *The Scientific Papers of James Clerk Maxwell*, ed. by W.D. Niven, Vol. 2 (The University Press, Cambridge 1890) p. 625
5.17 J.O. Hirschfelder, C.F. Curtiss, R.B. Bird: *Molecular Theory of Gases and Liquids* (Wiley, New York, Chapman and Hall, London 1954)
5.18 S. Chapman, T.G. Cowling: *The Mathematical Theory of Non-Uniform Gases*, 3rd ed. (The University Press, Cambridge 1970)
5.19 R. Fürth: Proc. Roy. Soc. London *A 179*, 461 (1942)
5.20 J. Stefan: Wien. Ber. *68*, 385 (1874)
5.21 J. Stefan: Wien. Ber. *83*, 943 (1881)
5.22 F.G. Volkov, A.M. Golovin: Zh. Prikl. Mekh. Tekh. Fiz. *1*, 78 (1970)
5.23 R.C. Reid, T.K. Sherwood: *The Properties of Gases and Liquids*, 2nd ed. (McGraw-Hill, New York 1966)

5.24 A. Wassiljewa: Phys. Z. *5*, 737 (1904)
5.25 A.L. Lindsay, L.A. Bromley: Ind. Eng. Chem. *42*, 1508 (1950)
5.26 E. Whalley: J. Chem. Phys. *19*, 509 (1951)
5.27 E.A. Mason, L. Mochick: in *Humidity and Moisture*, ed. by A. Wexler, Vol. 3 (Reinhold, New York, Chapman and Hall, London 1965) p. 257
5.28 J.L. Katz, P. Mirabel: J. Atmos. Sci. *32*, 646 (1975)
5.29 I. Langmuir: J. Amer. Chem. Soc. *37*, 417 (1915)
5.30 K. Schäfer: Z. Phys. *77*, 198 (1932)
5.31 N.A. Fuchs: Sov. Phys.-Tech. Phys. *3*, 140 (1958)
5.32 R.S. Bradley, M.G. Evans, R. Whytlaw-Gray: Proc. Roy. Soc. London *A 186*, 368 (1946)
5.33 P.G. Wright: Disc. Faraday Soc. *30*, 100 (1960)
5.34 J.C. Carstens, J.L. Kassner, Jr.: J. Rech. Atmos. *3*, 33 (1968)
5.35 N. Fukuta, L.A. Walter: J. Atmos. Sci. *27*, 1160 (1970)
5.36 J.C. Carstens, J. Podzimek, A. Saad: J. Atmos. Sci. *31*, 592 (1974)
5.37 B. Dahneke: private communication (1977)
5.38 P.L. Bhatnagar, E.P. Gross, M. Krook: Phys. Rev. *94*, 511 (1954)
5.39 J.R. Brock: J. Colloid Interface Sci. *22*, 513 (1966)
5.40 D.C. Sahni: J. Nucl. Energy A/B *20*, 915 (1966)
5.41 V.I. Smirnov: Pure Appl. Geophys. *86*, 184 (1971)
5.42 S.K. Loyalka: J. Chem. Phys. *58*, 354 (1973)
5.43 M.M.R. Williams: Z. Naturforsch. *30a*, 134 (1975)
5.44 P.N. Shankar: J. Fluid Mech. *40*, 385 (1970)
5.45 J. Jeans: *The Dynamical Theory of Gases* (Dover, New York 1954)
5.46 E.J. Davis, A.K. Ray: J. Aerosol Sci. *9*, 411 (1978)
5.47 N.A. Fuchs: in *Proceedings of the 7th Int. Conf. on Condensation and Ice Nuclei, Prague and Vienna*, ed. by J. Podzimek (Academia, Prague 1969) p. 10
5.48 B. Paul: ARS J. *32*, 1321 (1962)
5.49 G.M. Pound: J. Phys. Chem. Ref. Data *1*, 135 (1972)
5.50 T. Alty, C.A. Mackay: Proc. Roy. Soc. London *A 149*, 104 (1935)
5.51 K.C.D. Hickman: Ind. Eng. Chem. *46*, 1442 (1954)
5.52 A.F. Mills, R.A. Seban: Int. J. Heat Mass Transfer *10*, 1815 (1967)
5.53 L.J. Delaney, R.W. Houston, L.C. Eagleton: Chem. Eng. Sci. *19*, 105 (1964)
5.54 E. Akoy: "A Cloud Chamber Study for the Determination of Condensation Coefficient of Water"; M.S. Thesis, Brown University, Providence, Rhode Island, USA (1971)
5.55 A.M. Sinnarwalla, D.J. Alofs, J.C. Carstens: J. Atmos. Sci. *32*, 592 (1975)
5.56 U. Narusawa, G.S. Springer: J. Colloid Interface Sci. *50*, 392 (1975)
5.57 H. Reiss, V.K. La Mer: J. Chem. Phys. *18*, 1 (1950)
5.58 H. Reiss: J. Chem. Phys. *19*, 482 (1951)
5.59 W. Mordy: Tellus *11*, 16 (1959)
5.60 M. Neiburger, C.W. Chien: in *Geophysical Monograph Series, publ. by the American Geophysical Union*, ed. by H. Weickmann, W.E. Smith, Monograph 5 (Waverly, Baltimore 1960) p. 191
5.61 P.S. Brown, G. Arnason: J. Atmos. Sci. *30*, 245 (1973)
5.62 P.E. Wagner, F.G. Pohl: J. Colloid Interface Sci. *53*, 429 (1975)
5.63 J.C. Carstens, A. Williams, J.T. Zung: J. Atmos. Sci. *27*, 798 (1970)
5.64 A. Williams, J.C. Carstens: J. Atmos. Sci. *28*, 1298 (1971)
5.65 B.J. Mason: Proc. Phys. Soc. London *B 64*, 773 (1951)
5.66 R. Nestle: Z. Phys. *77*, 174 (1932)
5.67 J. Birks, R.S. Bradley: Proc. R. Soc. London *A 198*, 226 (1949)
5.68 R.S. Bradley, A.D. Shellard: Proc. R. Soc. London *A 198*, 239 (1949)
5.69 L. Monchick, H. Reiss: J. Chem. Phys. *22*, 831 (1954)
5.70 G.O. Langstroth, C.H.H. Diehl, E.J. Winhold: Can. J. Research *A 28*, 580 (1950)
5.71 E.J. Davis, E. Chorbajian: Ind. Eng. Chem. Fundam. *13*, 272 (1974)
5.72 R. Chang, E.J. Davis: J. Colloid Interface Sci. *54*, 352 (1976)
5.73 E.J. Davis, A.K. Ray: J. Chem. Phys. *67*, 414 (1977)
5.74 A.K. Ray, E.J. Davis, P. Ravindran: J. Chem. Phys. *71*, 582 (1979)
5.75 W.L. Dennis: Disc. Faraday Soc. *30*, 78 (1960)
5.76 A. Langsdorf: Rev. Sci. Instrum. *10*, 91 (1939)
5.77 A. Gagin, B. Terliuc: J. Rech. Atmos. *3*, 73 (1968)

5.78 J.P. Gollub, I. Chabay, W.H. Flygare: Appl. Opt. *12*, 2838 (1973)
5.79 J.P. Gollub, I. Chabay, W.H. Flygare: J. Chem. Phys. *61*, 2139 (1974)
5.80 N. Chodes, J. Warner, A. Gagin: J. Atmos. Sci. *31*, 1351 (1974)
5.81 M.A. Vietti, J.L. Fastook: J. Rech. Atmos. *9*, 181 (1975)
5.82 J.L. Katz, B.J. Ostermier: J. Chem. Phys. *47*, 478 (1967)
5.83 Coulier: J. Pharm. Chim. *(4) 22*, 165 (1875)
5.84 J. Aitken: Proc. Roy. Soc. Edinburgh *18*, 39 (1890/91)
5.85 C.T.R. Wilson: Proc. R. Soc. London *A 87*, 277 (1912)
5.86 C.T.R. Wilson: Proc. R. Soc. London *A 142*, 88 (1933)
5.87 J.G. Wilson: *The Principles of Cloud-Chamber Technique* (The University Press, Cambridge 1951)
5.88 C.W. Mettenburg, J.L. Kassner, Jr., D.A. Rinker: Bull. Am. Phys. Soc. (II) *3*, 302 (1958)
5.89 J.L. Kassner, Jr., J.C. Carstens, M.A. Vietti, A.H. Biermann, P.C.P. Yue, L.B. Allen, M.R. Eastburn, D.D. Hoffman, H.A. Noble, D.L. Packwood: J. Rech. Atmos. *3*, 45 (1968)
5.90 H. Israel, N. Nix: Z. Geophys. *32*, 175 (1966)
5.91 H. Israel, N. Nix: J. Rech. Atmos. *2*, 185 (1966)
5.92 R.G. Semonin, C.F. Hayes: J. Rech. Atmos. *3*, 287 (1968)
5.93 N. Nix: Arch. Meteorol. Geophys. Bioklimatol. *A 21*, 307 (1972)
5.94 N. Nix: "Kondensation und Verdampfung an künstlichen und natürlichen Aerosolen"; Ph.D. Thesis, Techn. Hochschule Aachen, FRG (1968)
5.95 J.L. Kassner, Jr., J.C. Carstens, L.B. Allen: J. Rech. Atmos. *3*, 25 (1968)
5.96 J.L. Kassner, Jr., J.C. Carstens, L.B. Allen: J. Atmos. Sci. *25*, 919 (1968)
5.97 N. Nix: Staub-Reinhalt. Luft *29*, 188 (1969)
5.98 T.G. Owe Berg, D.C. George: J. Geophys. Res. *73*, 3103 (1968)
5.99 B.G. Schuster, J.C. Carstens, J.L. Kassner, Jr.: J. Geophys. Res. *74*, 3447 (1969)
5.100 A. Cohen: Tellus *21*, 736 (1969)
5.101 P. Wagner: J. Colloid Interface Sci. *48*, 526 (1974)
5.102 M.A. Vietti, B.G. Schuster: J. Chem. Phys. *58*, 434 (1973)
5.103 M.A. Vietti, B.G. Schuster: J. Chem. Phys. *59*, 1499 (1973)
5.104 E.F. Allard, J.L. Kassner, Jr.: J. Chem. Phys. *42*, 1401 (1965)
5.105 J.C. Carstens, M.J. Carter: Proc. Int. Colloquium on Drops and Bubbles, ed. by D.J. Collins, M.S. Plesset, M.M. Saffren (U.S. Gov't. Printing Office 1976 - 685 - 197/2, Region No. 9-1, 1976) Vol. II, p. 529
5.106 J.C. Carstens, J.M. Carter, M.A. Vietti, B.G. Schuster: Trans. Am. Geophys. Union *54*, 286 (1973)
5.107 J.M. Carter, J.C. Carstens: Trans. Am. Geophys. Union *55*, 268 (1974)
5.108 P. Wagner: J. Colloid Interface Sci. *44*, 181 (1973)
5.109 P. Wagner: "Optische Bestimmung der Tröpfchenkonzentration in einer Expansionskammer bei Kondensation an verschiedenen Kernen", Proc. 2. Jahreskongress der Gesellschaft für Aerosolforschung, ed. by V. Böhlau, H. Straubel (Gesellschaft für Aerosolforschung, Bad Soden, West Germany 1974)
5.110 P.E. Wagner: J. Colloid Interface Sci. *53*, 439 (1975)
5.111 M.A. Vietti, J.L. Fastook: Trans. Am. Geophys. Union *55*, 268 (1974)
5.112 M.A. Vietti, J.L. Fastook: J. Chem. Phys. *65*, 174 (1976)
5.113 M.A. Vietti: Proc. Conf. on Cloud Physics and Atmospheric Electricity, ed. by P. Hobbs (American Meteorological Society, Boston, Massachusetts 1978) p. 60
5.114 P.E. Wagner, F.G. Pohl: "Dynamic Processes in Condensation Nuclei Counters - Model Calculations and Experimental Test", in *Atmospheric Aerosols and Nuclei*, ed. by A.F. Roddy, T.C. O'Connor (Galway University Press, Galway, Ireland 1981) p. 107
5.115 P.E. Wagner, F.G. Pohl: Gesellschaft Aerosolforsch. *5*, 279 (1977)
5.116 P.E. Wagner, F.G. Pohl: Staub-Reinhalt. Luft *38*, 72 (1978)
5.117 F. Pohl: "Zur Untersuchung der Wasserdampfkondensation in einer prozeßgesteuerten Expansionsnebelkammer und deren Anwendung zur Analyse des urbanen Aerosols"; Ph.D. Thesis, University of Vienna (1979)
5.118 M. Kerker: *The Scattering of Light and Other Electromagnetic Radiation* (Academic, New York 1969)

5.119 W. Szymanski, F.G. Pohl, P.E. Wagner: Gesellschaft Aerosolforsch. *7*, 285 (1979)
5.120 W. Szymanski, F.G. Pohl, P.E. Wagner: J. Aerosol Sci. *11*, 268 (1980)
5.121 P.E. Wagner, F.G. Pohl: Gesellschaft Aerosolforsch. *6*, 147 (1978)
5.122 P.E. Wagner, F.G. Pohl: J. Aerosol Sci. *10*, 204 (1979)

Additional References with Titles

Chapter 2

J. Barojas, E. Blaisten-Barojas, J. Flores: Two examples of electronic spectrum fluctuations in microparticles. Phys. Lett. A*69*, 142-144 (1978)

J. Barojas, E. Blaisten-Barojas, J. Flores: Electronic spectrum fluctuations in small particles. Kinam *1*, 361-381 (1979)

J. Barojas, E. Blaisten-Barojas, J. Flores: Strong coupling superconductivity in small particles. Kinam *2*, 71-77 (1980)

P.E. Batson: Damping of bulk plasmons in small aluminum spheres. Solid State Commun. *34*, 477-480 (1980)

R.E. Benner, P.W. Barber, J.F. Owen, R.K. Chang: Observation of structure resonances in the fluorescence spectra from microspheres. Phys. Rev. Lett. *44*, 475-478 (1980)

J. Buttet, J.-P. Borel (eds.): Second International Meeting on Small Particles and Inorganic Clusters. Ecole Polytechnique Fédérale, Lausanne, Switzerland, 8-12 Sept. 1980. Surface Science *106*, 1-608 (1981)

A.N. Chaba, R.K. Pathria: Evaluation of lattice sums using Poisson's summation formula IV. J. Phys. A*10*, 1823-1831 (1977)

S. Chippett, W.A. Gray: The size and optical properties of soot particles. Combust. Flame *31*, 149-159 (1978)

M. Cini: Classical and quantum aspects of size effects. J. Opt. Soc. Am. *71*, 386-392 (1981)

R. Fuchs: Optical excitations in small particles and thin films. J. Opt. Soc. Am. *71*, 379-382 (1981)

G.E. Galindo: "Absorcion de radiacion en polvos metalicos", Thesis, Universidad Nacional Autonoma de Mexico, Facultad de Ciencias (1981)

M. Hasgawa, K. Hoshino, M. Watabe: A theory of melting in metallic small particles. J. Phys. F*10*, 619-635 (1980)

A. Jay Palmer: Nonlinear optics in aerosols. Opt. Lett. *5*, 54-55 (1980)

K.L. Kliewer, R. Fuchs: "Theory of Dynamical Properties of Dielectric Surfaces", in *Advances in Chemical Physics*, ed. by I. Prigogine, S.A. Rice, Vol. 27 (Wiley, New York 1974) pp. 355-541

O. Matumura, M. Cho: Thermal-emission spectra of coagulated small MgO crystals. J. Opt. Soc. Am. *71*, 393-396 (1981)

K.A. Müller, M. Pomerantz, C.M. Knoedler: Inhomogeneous superconducting transitions in granular Al. Phys. Rev. Lett. *45*, 832-835 (1980)

G.A. Niklasson, C.G. Granqvist: Ultrafine nickel particles for photothermal conversion of solar energy. J. Appl. Phys. *50*, 5500-5505 (1979)

H.R. Pajkowski, R.J. Pathria: Criteria for the onset of Bose-Einstein condensation in ideal systems confined to restricted geometries. J. Phys. A*10*, 561-569 (1977)

F. Pu: Density of states, Poisson's formula of summation and Walfisz's formula. Phys. Lett. A*81*, 127-131 (1981)

K.H. Rieder, M. Ishigame, L. Genzel: Infrared absorption by coupled surface-phonon-surface-plasmon modes in microcrystals of CdO. Phys. Rev. B*6*, 3804-3810 (1972)

A. Schmitt-Ott, H.C. Siegmann: Photoelectron emission from small particles suspended in air. Appl. Phys. Lett. *32*, 710-713 (1978)

A. Schmitt-Ott, P. Schurtenberger, H.C. Siegmann: Enormous yield of photoelectrons from small particles. Phys. Rev. Lett. *45*, 1284-1287 (1980)

K. Weron: Thermodynamic properties of electrons in small metal particles. Acta Phys. Pol. A*51*, 323-333 (1980)

C.S. Zasada, R.K. Pathria: Low-temperature behavior of Bose systems confined to restricted geometries: Growth of the condensate fraction and spatial correlations. Phys. Rev. A*14*, 1269-1280 (1976)
C.S. Zasada, R.K. Pathria: Influence of boundary conditions on the growth of condensate fraction in a finite Bose system. Phys. Rev. A*15*, 2439-2443 (1977)
C.S. Zasada, R.K. Pathria: Statistical thermodynamics of superfluid helium confined to a cuboidal enclosure below 0.5 K. J. Phys. A*12*, 1531-1539 (1979)
C.S. Zasada, R.K. Pathria: Low-temperature thermodynamics of a weakly interacting Bose gas confined to restricted geometries. J. Math. Phys. *20*, 2559-2566 (1979)

Chapter 3

1. N.E. Brener, J.L. Fry: Hartree-Fock formalism for solids. I. Reciprocal-lattice expansion for the Hartree-Fock exchange term. Phys. Rev. B*17*, 506 (1978)
2. J.M. Andre, J. Delhalle, J.G. Fripiat, G. Leroy: Computation of LCAO wave functions for ground states of polymers and solids. Int. J. Quantum Chem. *5*, 67 (1971)
3. G. Grosso, G. Pastro Parravicini: Hartree-Fock energy bands by the orthogonalized-plane-wave method: Lithium hydride results. Phys. Rev. B*20*, 2366 (1979)
4. J.R. Smith, J.G. Gay, F.J. Arlinghaus: Self-consistent local-orbital method for calculating surface electronic structure: Application to Cu (100). Phys. Rev. B*21*, 2201 (1980)
5. J. Stöhr, L.I. Johansson, S. Brennan, M. Hecht, J.N. Miller: Surface extended x-ray-absorption-fine-structure study of oxygen interaction with Al (111) surfaces. Phys. Rev. B*22*, 4052 (1980)
6. D. Norman, S. Brennan, R. Jaeger, J. Stöhr: Structural models for the interaction of oxygen with Al (111) and Al implied by photoemission and surface EXAFS. Surf. Sci. *105*, L 297 (1981)
7. F. Soria, V. Martinez, M.C. Muñoz, J.L. Sacedón: Structure of the initial stages of oxidation of Al (111) surfaces from low-energy electron-diffraction and Auger electron spectroscopy. Phys. Rev. B*24*, 6926 (1981)
8. L. Kleinman, K. Mednick: Self-consistent calculations of oxygen monolayers on Al (111) films. Phys. Rev. B*23*, 4960 (1981)

Subject Index

Inverse Scattering Problems

in Optics

Editor: **H.P. Baltes**
With a Foreword by R. Jost

1980. 49 figures, 2 tables. XIV, 313 pages
(Topics in Current Physics, Volume 20)
ISBN 3-540-10104-7

Contents:
H.P. Baltes: Progress in Inverse Optical Problems. – *G. Ross, M.A. Fiddy, M. Nieto-Vesperinas:* The Inverse Scattering Problem in Structural Determinations. – *E. Jakeman, P.N. Pusey:* Photon-Counting Statistics of Optical Scintillation. – *A. Selloni:* Microscopic Models of Photodetection. – *M. Bertero, C. De Mol, G.A. Viano:* The Stability of Inverse Problems. – *R. Goulard, P.J. Emmerman:* Combustion Diagnostics by Multi-angular Absorption. – *W.-M. Boerner:* Polarization Utilization in Electromagnetic Inverse Scattering.

Inverse Source Problems in Optics

Editor: **H.P. Baltes**
With a Foreword by J.-F. Moser

1978. 32 figures. XI, 204 pages. (Topics in Current Physics, Volume 9)
ISBN 3-540-09021-5

Contents:
H.P. Baltes: Introduction. – *H.A. Ferwerda:* The Phase Reconstruction Problem for Wave Amplitudes and Coherence Functions. – *B.J. Hoenders:* The Uniqueness of Inverse Problems. – *H.G. Schmidt-Weinmar:* Spatial Resolution of Subwavelength Sources from Optical Far-Zone Data. – *H.P. Baltes, J. Geist, A. Walther:* Radiometry and Coherence. – *A. Zardecki:* Statistical Features of Phase Screens from Scattering Data.

Laser Monitoring of the Atmosphere

Editor: **E.D. Hinkley**

1976. 84 figures. XV, 380 pages
(Topics in Applied Physics, Volume 14)
ISBN 3-540-07743-X

Contents:
E.D. Hinkley: Introduction. – *S.H. Melfi:* Remote Sensing for Air Quality Management. – *V.E. Zuev:* Laser-Light Transmission Through the Atmosphere. – *R.T.H. Collis, P.B. Russel:* Lidar Measurement of Particles and Gases by Elastic Back scattering and Differential Absorption. – *H. Inaba:* Detection of Atoms and Molecules by Raman Scattering and Resonance Fluorescence. – *E.D. Hinkley, R.T. Ku, P.L. Kelley:* Techniques for Detection of Molecular Pollutants by Absorption of Laser Radiation. – *R.T. Menzies:* Laser Heterodyne Detection Techniques.

Theory of Chemisorption

Editor: **J.R. Smith**

1980. 116 figures, 8 tables. XI, 240 pages
(Topics in Current Physics, Volume 19)
ISBN 3-540-09891-7

Contents:
J.R. Smith: Introduction. – *S.C. Ying:* Density Functional Theory of Chemisorption on Simple Metals. – *J.A. Appelbaum, D.R. Hamann:* Chemisorption on Semiconductor Surfaces. – *F.J. Arlinghaus, J.G. Gay, J.R. Smith:* Chemisorption on d-Band Metals. – *B. Kunz:* Cluster Chemisorption. – *T. Wolfram, S. Elliatioğlu:* Concepts of Surface States and Chemisorption and d-Band Perovskites. – *T.L. Einstein, J.A. Hertz, J.R. Schrieffer:* Theoretical Issues in Chemisorption.

Springer-Verlag Berlin Heidelberg New York